# Occupational Safety and Health *Simplified* for the Food Manufacturing Industry

FRANK R. SPELLMAN and REVONNA M. BIEBER

GOVERNMENT INSTITUTES
An imprint of
The Scarecrow Press, Inc.
*Lanham, Maryland • Toronto • Plymouth, UK*
2008

**Government Institutes**

Published in the United States of America
by Government Institutes, an imprint of The Scarecrow Press, Inc.
A wholly owned subsidiary of
The Rowman & Littlefield Publishing Group, Inc.
4501 Forbes Boulevard, Suite 200
Lanham, Maryland 20706
http://www.govinstpress.com/

Estover Road
Plymouth PL6 7PY
United Kingdom

The reader should not rely on this publication to address specific questions that apply to a
particular set of facts. The author and the publisher make no representation or warranty, express or
implied, as to the completeness, correctness, or utility of the information in this publication. In
addition, the author and the publisher assume no liability of any kind whatsoever resulting from
the use of or reliance upon the contents of this book.

British Library Cataloguing in Publication Information Available

**Library of Congress Cataloging-in-Publication Data**

Spellman, Frank R.
  Occupational safety and health simplified for the food manufacturing industry / Frank R.
Spellman & Revonna M. Bieber.
       p. ; cm.
  Includes bibliographical references and index.
  ISBN-13: 978-0-86587-184-7 (pbk. : alk. paper)
  ISBN-10: 0-86587-184-1 (pbk. : alk. paper)
  eISBN-13: 978-1-60590-270-8
  eISBN-10: 1-60590-270-5
  1. Food industry and trade—United States—Safety measures. 2. Food industry and trade—
Health aspects—United States. 3. United States. Occupational Safety and Health Administration. I.
Bieber, Revonna M., 1976– II. Title.
  [DNLM: 1. United States. Occupational Safety and Health Administration. 2. Food Industry—
standards—United States. 3. Occupational Health—United States. 4. Occupational Exposure—
prevention & control—United States. 5. Safety Management—standards—United States. WA 400
S743o 2008]
  RC965.F58S64 2008
  363.19'260973—dc22                                                      2008010068

∞™ The paper used in this publication meets the minimum requirements of
American National Standard for Information Sciences—Permanence of
Paper for Printed Library Materials, ANSI/NISO Z39.48-1992.
Manufactured in the United States of America.

# Contents

# Preface

The second of a new Government Institutes series, *Occupational Safety and Health* Simplified *for the Food Manufacturing Industry* is a reference that serves food manufacturing businesses and managers who want quick answers to complicated questions—to help employers and employees handle the safety hazards they deal with on a daily basis. This book is an effort to simplify in a single volume everything that most employers in the food manufacturing industry need to know about applicable Occupational Safety and Health Administration (OSHA) standards. Moreover, because it is a reference book, this ready handbook is designed to be consulted frequently; to be taken to job locations; to help the user understand and make sense of complicated OSHA regulations. Because this text is written in plain English and includes applicable/appropriate hazard information highlighted in Spanish, the process of understanding is simplified. However, it is important to note that it is a starting point only. Compliance with the law can only be achieved by consulting the actual Code of Federal Regulations (CFR)—in this case, 29 CFR 1910 General Industry Standards (the OSHA "bible") and 29 CFR 1928 Agricultural Standards. While this book is intended to help you make sense of the regulations, you must go to the regulations themselves in order to ensure compliance with the law—and, more importantly, to protect employees.

*Occupational Safety and Health* Simplified *for the Food Manufacturing Industry*

- provides farm-to-fork coverage (food starts its journey at the farm and eventually is transported, making it accessible to us)
- covers how to prevent accidents from the most serious hazards in the food manufacturing industry, such as
  - respiratory hazards from inhalation of flavorings
  - electrical incidents (includes new OSHA/NFPA 70E requirements)
  - slips, trips, and falls

- cuts (lacerations)
- ergonomic hazards
- confined space entry hazards
- stored energy hazards
- material handling hazards
- machine guarding
- wet/cold environments
- infectious disease
- amputations
- radiation hazards
- noise hazards
- chemical hazards
- dermatitis
- simplifies all of OSHA's Food Manufacturing Standards Requirements
- simplifies all of OSHA's Food Manufacturing Training Requirements
- simplifies how to communicate effectively with Spanish-speaking employees on your job site
- includes a comprehensive, easy-to-use checklist to ensure regulatory compliance and to conduct safety inspections of food manufacturing site
- includes a sample food manufacturing safety program to meet OSHA compliance requirements

This book is written to help businesses that need the most current information in the food manufacturing industry and should be used as a reference book for quick answers to complicated questions.

It is the authors' hope that this book will help employers prevent many of the food manufacturing injuries and fatalities that occur each year, while simplifying the OSHA compliance process. However, it is very important to remember that the only way to ensure compliance with the law is to consult and comply with the rules in the Code of Federal Regulations. The applicable volumes of the CFR can be obtained from Government Institutes or the Government Printing Office.

# 1

# Introduction

OSHA News Release [12/15/2003]
**OSHA Cites Austin Food Manufacturing Company for Worker Death**
*Michael Angelo's Gourmet Foods Inc. Will Pay $140,220 Fine*

**Austin, Tex.**—Michael Angelo's Gourmet Foods Inc. in Austin, Texas, has agreed to pay $140,220 in penalties for citations issued by the U.S. Department of Labor's Occupational Safety and Health Administration (OSHA) for failure to provide employees with adequate protection and training to prevent machinery from starting up during cleaning operations.

A worker for the frozen food manufacturer was killed in June when he was pulled into a meat mixer.

OSHA began its investigation on June 13 after the employee was found dead by a co-worker at the company's headquarters. . . . Michael Angelo Gourmet Foods employs about 440 workers at that location.

The company was cited for 11 safety and health violations for exposing employees to electrical hazards such as defective electrical cords, failing to properly guard machinery that could cause amputations and other injuries, failing to ensure locks were provided on machinery requiring maintenance, and failing to train its employees to perform lockout procedures. Lockout/tag-out involves shutting off and locking out the energy source to a machine that is undergoing maintenance or repair to prevent an accidental startup.

OSHA has inspected Michael Angelo twice since 2000, and the company has paid more than $63,000 in penalties for similar violations.

Michael Angelo's Gourmet Foods has agreed to fully comply with OSHA's standards by providing lockout/tag-out training for its employees, managers and supervisors. The company also has agreed to abate the violations and certify their abatement within 10 days. In addition, the company will retain the services of an outside consultant to review its safety and health program. (OSHA 2003)

## NATURE OF THE INDUSTRY

DID YOU KNOW?

Food manufacturing has one of the highest incidences of injury and illness among all industries; animal slaughtering plants have the highest incidence among all food manufacturing industries. (BLS 2008)

As the Bureau of Labor Statistics points out,

workers in the food manufacturing industry link farmers and other agricultural products with consumers. They do this by processing raw fruits, vegetables, grains, meats, and dairy products into finished goods ready for the grocer or wholesaler to sell to households, restaurants, or institutional food services.

Food manufacturing workers perform tasks as varied as the many foods we eat. For example, they slaughter, dress, and cut meat or poultry; process milk, cheese, and other dairy products; can and preserve fruits, vegetables, and frozen specialties; manufacture flour, cereal, pet foods, and other grain mill products; make bread, cookies, cakes, and other bakery products; manufacture sugar and candy and other confectionery products; process shortening, margarine, and other fats and oils; and prepare packaged seafood, coffee, potato and corn chips, and peanut butter. Although this list is long, it is not exhaustive: Food manufacturing workers also play a part in delivering numerous other food products to our tables.

Quality control and quality assurance (QC & QA) are vital to this industry. The U.S. Department of Agriculture's (USDA) Food Safety and Inspection Service branch oversees all aspects of food manufacturing. In addition, other food safety programs have been adopted recently as issues of chemical contamination and the growing number of new food-borne pathogens remains a public health concern. For example, by applying science-based controls from raw materials to finished products, a food safety program called Hazard Analysis and Critical Control Point (HACCP) focuses on identifying hazards and preventing them from contaminating food in early stages of meat processing (BLS 2008).

HACCP is an evaluation system to identify, monitor, and control contamination risks in food manufacturing establishments. Note that General Mills, one of the leaders in HACCP implementation, defines a "critical hazard" as an imminent health hazard or a guest dissatisfaction. Since incorporating this system totally into its operations, General Mills has increased food quality substantially.

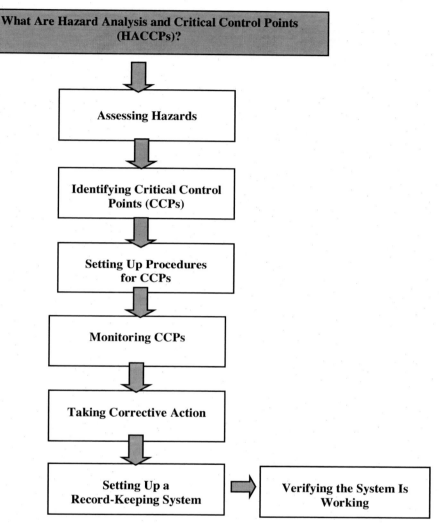

FIGURE 1
Hazard Analysis and Critical and Control Points (HACCPs).

DID YOU KNOW?

In food manufacturing, production workers account for 53 percent of all jobs. (BLS 2008)

Table 1.1. Distribution of wage and salary employment in food manufacturing by industry segment, 2006 (Employment in thousands)

| Industry Segment | 2006 Employment | 2006 Percent Change |
|---|---|---|
| Food manufacturing, total | 1,484 | 0.3 |
| Animal slaughtering and processing | 509 | 11.8 |
| Bakeries and tortilla manufacturing | 280 | 1.5 |
| Fruit and vegetable preserving and specialty food manufacturing | 177 | −12.3 |
| Dairy product manufacturing | 132 | −3.9 |
| Sugar and confectionery product manufacturing | 75 | −18.0 |
| Grain and oilseed milling | 60 | −15.0 |
| Animal food manufacturing | 50 | −15.5 |
| Seafood product preparation and packaging | 40 | −11.0 |
| Other food manufacturing | 160 | 1.6 |

Source: BLS 2008.

About 34 percent of all food manufacturing workers are employed in plants that slaughter and process animals, and another 19 percent work in establishments that make bakery goods (see table 1.1). Seafood product preparation and packaging, the smallest sector of the food manufacturing industry, accounts for only 3 percent of all jobs.

## FARM-TO-FORK CONTINUUM: AGRICULTURE

DID YOU KNOW?

In any given year, more than 700 Americans die in farm-related accidents, with children accounting for between 100 and 300 of those deaths.

Although this book is not intended to be a detailed description and accounting of agricultural safety and safe work practices, we feel it is important to include a brief overview of farm safety in this discussion.

Some of us might imagine living and working on a farm in the stereotypical mode portrayed by Grant Wood in his classic painting *American Gothic* (1930), in which a pitchfork-holding Midwestern farmer stands next to his daughter before a large Gothic window. Most of us probably imagine living and working on a farm out in the country as a safe, healthy, quiet, peaceful, outdoorsy, get-back-to-nature activity; for the most part this is the case.

Most people in risky occupations, including farming, live to a ripe old age. However, like other risky occupations, farming has many hazards that the average person has no knowledge or full appreciation of. This lack of knowledge of potential safety problems can be extremely costly, costing fingers, hands, arms, and/or lives. Another unavoidable potentially dangerous circumstance in farming has to do with the occupation's need for farmers to perform many solitary tasks way out in the back forty, so to speak. Working alone, far from other people who could help in an emergency, adds to the severity of many accidents.

## DID YOU KNOW?

In any given year, the number of farm fatalities is usually exceeded only by those related to mining.

Contrary to the popular image of fresh air and peaceful surroundings, a farm is not a hazard-free work setting. [As mentioned above,] every year thousands of farm workers are injured and hundreds more die in farming accidents. According to the National Safety Council, agriculture is the most hazardous industry in the nation.

Farm workers—including farm families and migrant workers—are exposed to hazards such as the following:

- chemicals/pesticides
- cold
- dust
- electricity
- grain bins
- hand tools
- highway traffic
- lifting
- livestock handling
- machinery/equipment
- manure pits
- mud
- noise
- ponds
- silos
- slips/trips/falls
- sun/heat
- toxic gases
- tractors
- wells

**High-Risk Factors on Farms**

The following factors may increase risk of injury or illness for farm workers:

- **Age:** Injury rates are highest among children age 15 and under and adults over 65.
- **Equipment and Machinery:** Most farm accidents and fatalities involve machinery. Proper machine guarding and doing equipment maintenance according to manufacturers' recommendations can help prevent accidents.
- **Protective Equipment:** Using protective equipment, such as seat belts on tractors, and personal protective equipment (such as safety gloves, coveralls, boots, hats, aprons, goggles, and face shields) could significantly reduce farming injuries.
- **Medical Care:** Hospitals and emergency medical care are typically not readily accessible in rural areas near farms.

**How Can We Improve Farm Safety?**

OSHA (2007) states that farmers can start by increasing their awareness of farming hazards and making a conscious effort to prepare for emergency situations including fires, vehicle accidents, electrical shocks from equipment and wires, and chemical exposures. Farmers should be especially alert to hazards that may affect children and the elderly. Hazards should be minimized by careful selection of good tools and equipment. Seat belts should always be used when operating tractors, and good housekeeping practices should be established and maintained. Here are some other steps that can be taken to reduce illnesses and injuries on the farm:

- Reading and following instructions in equipment operator's manuals and on product labels.
- Inspecting equipment routinely for problems that may cause accidents.
- Discussing safety hazards and emergency procedures with farm workers.
- Installing approved rollover protective structures, protective enclosures, or protective frames on tractors.
- Making sure that guards on farm equipment are replaced after maintenance.
- Reviewing and following instructions in material safety data sheets (MSDSs) and on labels that come with chemical products, and communicating information on these hazards to farm workers.
- Taking precautions to prevent entrapment and suffocation caused by unstable surfaces of grain storage bins, silos, or hoppers. (It is prohibited by OSHA to "walk the grain.")
- Being aware that methane gas, carbon dioxide, ammonia, and hydrogen sulfide can form in unventilated grain silos and manure pits and can suffocate or poison workers or explode.
- Taking advantage of safety equipment, such as bypass starter covers, power take-off master shields, and slow-moving vehicle emblems (OSHA 2007a).

DID YOU KNOW?

"Walking down grain" and similar practices where employees walk on grain to get grain to flow out of a grain storage structure or where employees are on moving grain are prohibited by 29 CFR 1910.272(g)(1)(iv).

### The 1928 Agriculture Standard

If federal OSHA administers your state and you employ workers outside your immediate family, you must have a copy of the appropriate volume or volumes of the Code of Federal Regulations, 29 CFR 1910 (General Industry Standards) and 29 CFR 1928 (Agriculture Standards). There must be a copy physically present at locations where people you hire work (see table 1.2).

**Table 1.2.  29 CFR standards**

## 29 CFR 1928—Agriculture Standards

**Subpart C—Employee Operating Instruction**

| Section | Designation |
|---|---|
| 1928.51 | Roll-over protective structures (ROPs) for tractors used in agricultural operations. |
| 1928.52 | Protective frames for wheel-type agricultural tractors—test procedures and performance requirements. |
| 1928.53 | Protective enclosures for wheel-type agricultural tractors—test procedures and performance requirements. |

**Subpart D—Safety for Agricultural Equipment**

| Section | Designation |
|---|---|
| 1928.57 | Guarding of farm field equipment, farmstead equipment, and cotton gins. |

**Subpart I—General Environment Controls**

| Section | Designation |
|---|---|
| 1928.110 | Field sanitation |

**Subpart M—Occupational Health**

| | |
|---|---|
| 1928.1027 | Cadmium |

## 29 CFR 1910—General Industry Standards

**Subpart J—General Environmental Controls**

| | |
|---|---|
| 1910.142 | Temporary labor camps |
| 1910.145 | Specifications for accident prevention signs and tags |

**Subpart H—Hazardous Materials**

| | |
|---|---|
| 1910.111 | Storage and handling of anhydrous ammonia |

**Table 1.2.** (*Continued*)

**Subpart R—Special Industries**

| 1910.266 | Logging operations |
|----------|--------------------|

**Subpart Z—Toxic and Hazardous Substances**

| 1910.1200 | Hazard communication |
|-----------|---------------------|
| 1910.1201 | Retention of DOT markings, placards, and labels |
| 1910.1027 | Cadmium |

DID YOU KNOW?

Most food production jobs require little formal education or training. Many can be learned in a few days. (BLS 2008)

## FOOD MANUFACTURING WORKING CONDITIONS

Many production jobs in food manufacturing involve repetitive, physically demanding work. Food manufacturing workers are highly susceptible to repetitive-strain injuries to their hands, wrists, and elbows. This type of injury is especially common in meat-processing and poultry-processing plants. Production workers often stand for long periods and may be required to lift heavy objects or use cutting, slicing, grinding, and other dangerous tools and machines. To deal with difficult working conditions, ergonomic programs have been introduced to cut down on work-related accidents and injuries.

In 2006, there were 7.4 cases of work-related injury or illness per 100 full-time food manufacturing workers, *much higher than the rate of 4.4 cases for the private sector as a whole.* Injury rates vary significantly among specific food manufacturing industries, ranging from a low of 1.8 per 100 workers in retail bakeries to 12.5 per 100 in animal slaughtering plants, the highest rate in food manufacturing (BLS 2008).

DID YOU KNOW?

Eighty-nine percent of establishments in food manufacturing employ fewer than 100 workers. (BLS 2008)

In an effort to reduce occupational hazards, many plants have redesigned equipment, increased the use of job rotation, allowed longer or more frequent breaks, and developed training programs in safe work practices. Furthermore, meat and poultry plants must comply with a wide array of Occupational Safety and Health Administration (OSHA) regulations ensuring a safe work environment. Although injury rates remain high, training and other changes have reduced those rates. Some workers wear protective hats or masks, gloves, aprons, and shoes. In many companies, uniforms and protective clothing are changed daily for reasons of sanitation.

Because of the considerable mechanization in the industry, most food manufacturing plants are noisy, with limited opportunities for interaction among workers. In some highly automated plants, "hands-on" manual work has been replaced by computers and factory automation, resulting in less waste and higher productivity. While much of the basic production—such as trimming, chopping, and sorting— will remain labor intensive for many years to come, automation is increasingly being applied to various functions including inventory management, product movement, and quality control issues such as packing and inspection (BLS 2008).

### DID YOU KNOW?

Automation and increasing productivity will limit employment growth, but unlike many other industries, food manufacturing is not highly sensitive to economic conditions. (BLS 2008)

Working conditions also depend on the type of food being processed. For example, some bakery employees work at night or on weekends and spend much of their shifts near ovens that can be uncomfortably hot. In contrast, workers in dairies and meat-processing plants typically work daylight hours and may experience cold and damp conditions. Some plants, such as those producing processed fruits and vegetables, operate on a seasonal basis, so workers are not guaranteed steady, year-round employment and occasionally travel from region to region seeking work. These plants are increasingly rare, however, as the industry continues to diversify and manufacturing plants produce alternative foods during otherwise inactive periods (BLS 2008).

## FOOD MANUFACTURING EMPLOYMENT

In 2006, the food manufacturing industry provided 1.5 million jobs. Almost all employees were wage and salary workers; only a few were self-employed and unpaid family workers. In 2006, about 28,000 establishments manufactured food, with 89 percent employing fewer than 100 workers. Nevertheless, establishments employing 500 or more workers accounted for 36 percent of all jobs.

The employment distribution in this industry varies widely. Animal slaughtering and processing employs the largest proportion of workers. Economic changes in livestock farming and slaughtering plants have changed the industry. Increasingly, fewer but larger farms are producing the vast majority of livestock in the United States. Similarly, there are now fewer, but much larger, meat-processing plants, owned by fewer companies—a development that has tended to concentrate employment in a few locations.

Food manufacturing workers are found in all states, although some sectors of the industry are concentrated in certain parts of the country. For example, in 2006, California, Illinois, Iowa, Pennsylvania, and Texas employed 24 percent of all workers in animal slaughtering and processing. That same year, Wisconsin employed 32 percent of all cheese manufacturing workers, and California accounted for 24 percent of fruit and vegetable canning, pickling, and drying workers (BLS 2008).

## OCCUPATIONS IN THE FOOD MANUFACTURING INDUSTRY

The food manufacturing industry employs many different types of workers. More than half, or 54 percent, are production workers, including skilled precision workers and less skilled machine operators and laborers. Production jobs require manual dexterity, good hand-eye coordination, and, in some sectors of the industry, strength.

### Meat, Poultry, and Fish Production Workers

Red-meat production is the most labor-intensive food processing operation. Animals are not uniform in size, and *slaughterers and meatpackers* must slaughter, skin, eviscerate, and cut each carcass into large pieces. They usually do this work by hand, using large, suspended power saws. Increasingly, most plants today require slaughterers and meat packers to further process the large parts by cleaning, salting, and cutting them into tenders and chucks to make them readily available for retail use. Such prepackaged meat products are increasingly preferred by retailers and grocers as they can be easily displayed and sold without the need of a butcher.

*Meat, poultry, and fish cutters and trimmers* use hand tools to break down the large primary cuts into smaller sizes for shipment to wholesalers and retailers. Such ready-to-cook meat products are increasingly prepared at processing plants where preparation may now entail filleting; cutting into bite-sized pieces or tenders; preparing and adding vegetables; and applying sauces and flavorings, marinades, or breading. These workers use knives and other hand tools for these processes (BLS 2008).

Most butchers and meat, poultry, and fish cutters and trimmers frequently work in cold, damp rooms that are refrigerated to prevent meat from spoiling. Workrooms are often damp because meat cutting generates large amounts of blood and condensation. These occupations require physical strength to lift and carry large cuts of meat and the ability to stand for long periods. Butchers and meat, poultry, and fish cutters and trimmers work in clean and sanitary conditions; however, their clothing is often soiled with animal blood and the air may smell unpleasant. They work with powerful cutting equipment and are susceptible to cuts on the fingers or hands. Risks are minimized with the proper use of equipment and hand and stomach guards. The repetitive nature of the work, such as cutting and slicing, may lead to wrist damage (carpal tunnel syndrome) (CALMIS 2007).

### Bakers

*Bakers* mix and bake ingredients according to recipes to produce breads, cakes, pastries, and other goods. Bakers produce goods in large quantities, using mixing machines, ovens, and other equipment (BLS 2008).

DID YOU KNOW?

On January 18, 2007, a wide range of safety and health hazards at a Syracuse bakery resulted in $120,600 in proposed fines from OSHA. Penny Curtiss Baking Co. Inc., which manufactures bread and other bakery products, was cited for a total of 42 alleged serious safety and health hazards at its production plant, following an OSHA inspection begun in July in response to an employee complaint. (OSHA 2007b)

Well-managed bakeries are generally kept spotlessly clean, and personal cleanliness is very important. However, work areas can be uncomfortably hot and noisy. Many employers who require uniforms furnish and launder employee uniforms. Oven mitts are also usually supplied to employees when necessary. Bakery production jobs are usually performed at a fast, steady pace while standing. Many plant jobs involve strenuous physical work, including heavy lifting, despite the use of machinery.

### Hands-On Food Production Workers

Many food manufacturing workers use their hands or small hand tools to do their jobs. *Cannery workers* perform a variety of routine tasks—such as sorting, grading, washing, trimming, peeling, or slicing—in the canning, freezing, or packing of food products. *Hand food decorators* apply artistic touches to prepared foods. *Candy molders* and *marzipan (confection) shapers* form sweets into fancy shapes by hand.

### DID YOU KNOW?

On March 11, 2004, because of a fatality at a Mississippi cannery, OSHA proposed penalties totaling $229,000. (OSHA 2004)

### Food Manufacturing Machine Operators

With increasing levels of automation in the food manufacturing industry, a growing number of workers are operating machines. For example, *food batchmakers* operate equipment that mixes, blends, or cooks ingredients used in manufacturing various foods, such as cheese, candy, honey, and tomato sauce. *Dairy processing equipment operators* process milk, cream, cheese, and other dairy products. *Cutting and slicing machine operators* slice bacon, bread, cheese, and other foods. *Mixing and blending machine operators* produce dough, batter, fruit juices, or spices. *Crushing and grinding machine operators* turn raw grains into cereals, flour, and other milled-grain products, and they produce oils from nuts or seeds. *Extruding and forming machine operators* produce molded food and candy, and *casing finishers and stuffers* make sausage links and similar products. *Bottle packers and bottle fillers* operate machines that fill bottles and jars with preserves, pickles, and other foodstuffs.

### Food Cooking Machine Operators

*Food cooking machine operators and tenders* steam, deep-fry, boil, or pressure-cook meats, grains, sugar, cheese, or vegetables. *Food and tobacco roasting, baking, and drying machine operators and tenders* operate equipment that roasts grains, nuts, or coffee beans and tend ovens, kilns, dryers, and other equipment that removes moisture from macaroni, coffee beans, cocoa, and grain. *Baking equipment operators* tend ovens that bake bread, pastries, and other products. Some foods—ice cream, frozen specialties, and meat, for example—are placed in freezers or refrigerators by *cooling and freezing equipment operators*. Other workers tend machines and equipment that clean and wash food or food-processing equipment. Some machine operators also clean and maintain machines and perform duties such as checking the weight of foods.

### Food Manufacturing Maintenance Workers

Many other workers are needed to keep food manufacturing plants and equipment in good working order. *Industrial machinery mechanics* repair and maintain production machines and equipment. *Maintenance repairers* perform routine maintenance on machinery, such as changing and lubricating parts. Specialized mechanics include *heating, air-conditioning, and refrigeration mechanics, farm equipment mechanics*, and *diesel engine specialists*.

### QA/QC

Still other workers directly oversee the quality of the work and of final products. *Supervisors* direct the activities of production workers. *Graders and sorters* of agricultural products, *production inspectors*, and *quality control technicians* evaluate foodstuffs before, during, or after processing.

### Packagers/Transporters

Food may spoil if not packaged properly or delivered promptly, so packaging and transportation employees play a vital role in the industry. Among those are *freight, stock, and material movers*, who manually move materials; *hand packers and packagers*, who pack bottles and other items as they come off the production line; and *machine feeders and offbearers*, who feed materials into machines and remove goods from the end of the production line. *Industrial truck and tractor operators* drive gasoline- or electric-powered vehicles equipped with forklifts, elevated platforms, or trailer hitches to move goods around a storage facility. *Truck drivers* transport and deliver livestock, materials, or merchandise and may load and unload trucks. *Driver/sales workers* drive company vehicles over established routes to deliver and sell goods, such as bakery items, beverages, and vending-machine products.

### Managers

The food manufacturing industry also employs a variety of managerial and professional workers. *Managers* include top executives, who make policy decisions; *industrial production managers*, who organize, direct, and control the operation of the manufacturing plant; and *advertising, marketing, promotions, public relations, and sales managers*, who direct advertising, sales promotion, and community relations programs.

### Professional Staff

Engineers, scientists, and technicians are becoming increasingly important as the food manufacturing industry implements new automation and food safety processes. These workers include *industrial engineers*, who plan equipment layout and workflow in manufacturing plants, emphasizing efficiency and safety. Also, *mechanical engineers* plan, design, and oversee the installation of tools, equipment, and machines. *Chemists* perform tests to develop new products and maintain the quality of existing products. *Computer programmers and systems analysts* develop computer systems and programs to support management and scientific research. *Food scientists and technologists* work in research laboratories or on production lines to develop new products, test current ones, and control food quality, including minimizing food-borne pathogens.

### Sales/Marketing Workers

Finally, many sales workers, including *sales representatives, wholesale and manufacturing*, are needed to sell the manufactured goods to wholesale and retail establishments. *Bookkeeping, accounting, and auditing clerks and procurement clerks* keep track of the food products going into and out of the plant. *Janitors and cleaners* keep buildings clean and orderly (BLS 2008).

DID YOU KNOW?

The injury and illness rate for the food manufacturing industry is significantly higher than that for the manufacturing sector as a whole.

## FOOD MANUFACTURING TRAINING

Most production jobs in food manufacturing require little formal education. Graduation from high school is preferred, but not always required. In general, inexperienced workers start as helpers to experienced workers and learn skills on the job. Many of these entry-level jobs can be learned in a few days. Typical jobs include operating a bread-slicing machine, washing fruits and vegetables before processing begins, hauling carcasses, and packing bottles as they come off the production line. Even though it may not take long to learn to operate a piece of equipment, employees may need several years of experience to enable them to keep the equipment running smoothly, efficiently, and safely.

Some food manufacturing workers need specialized training and education. Inspectors and quality control workers, for example, are trained in food safety and usually need a certificate to be employed in a food manufacturing plant. Often, USDA-appointed plant inspectors possess a bachelor's degree in agricultural or food science. Formal educational requirements for managers in food manufacturing plants range from two-year degrees to master's degrees. Those who hold research positions, such as food scientists, usually need a master's or doctoral degree.

In addition to participating in specialized training, a growing number of workers receive broader training to perform a number of jobs. The need for flexibility in more automated workplaces has meant that many food manufacturing workers are learning new tasks and being trained to work effectively in teams. Some specialized training is provided for bakers and some other positions (BLS 2008).

## REFERENCES AND RECOMMENDED READING

Bureau of Labor Statistics (BLS). 2008. Career Guide to Industries: Food Manufacturing. www.bls.gov/oco/cg/cgs011.htm (accessed March 20, 2008).

California Labor Market Information Services (CALMIS). 2007. Food Manufacturing. www.caljobs.ca.gov (accessed July 30, 2007).

National Safety Council. 2007. Farm Facts. www.nsc.org/farmsafe/facts.htm (accessed July 28, 2007).

Occupational Safety and Health Administration (OSHA). 2003. OSHA Cites Austin Food Manufacturing Company for Worker Death. www.dol.gov/opa/media/press/osha/OSHA2003904.htm (accessed July 27, 2007).

———. 2004. Fatality at Mississippi Cannery Leads to OSHA Citations: Agency Proposes Penalties Totaling $229,000. http://www.osha.gov/pls/oshaweb/owadisp.show_document?p_table=NEWS_RELEASES&p_id=10731 (accessed July 27, 2007).

———. 2007a. OSHA Fact Sheet: Farm Safety. www.osha.gov/OshDoc/data_General_Facts/FarmFactS2.pdf (accessed July 27, 2007).

———. 2007b. U.S. Labor Department's OSHA Fines Penny Curtiss Baking Co. $120,600 for Safety and Health Hazards at Syracuse Bakery. http://www.osha.gov/pls/oshaweb/owadisp.show_document?p_table=NEWS_RELEASES&p_id=13518 (accessed July 27, 2007).

# 2

# The OSH Act
# and Food Manufacturing

**OSHA Fines Granny's**

On Monday, October 3, 2005, in Frankfort, N.Y., OSHA levied $142,200 in fines against Granny's Kitchens. The doughnut manufacturer was cited for 20 alleged willful and serious violations of workplace safety standards at its Frankfort production plant. (OSHA 2005)

Based on personal experience, we have found that the attitude expressed in the box below is rather common in industry. However, we have also found that companies that are in compliance with the regulations of OSHA and other regulators (e.g., DOT, EPA) usually have little to fear from OSHA.

## OSHA AT THE DOOR!

Costly modifications of existing installations to meet new legal demands is not only a headache-generator for any facility manager, it also can be very costly to the company in terms of both money and workers' time. Companies are in business, obviously, to make money—to maintain or (we hope) improve their bottom line. The last thing any manager who is fighting competition and other costly impediments to making his or her company profitable is to have "those briefcase-carrying so-and-sos coming into my plant and telling ME I have to have this and I have to have that . . . or else!" Not only do they waste my time, but they also make me spend money on things that cut profits—things that don't contribute to the bottom line. (Spellman 1998)

### OCCUPATIONAL SAFETY AND HEALTH ACT (1970)

While some federal safety legislation was passed prior to 1970, this legislation only affected a small fraction of the U.S. workforce. At the end of the 1960s, two shortcomings became blatantly obvious: (1) a new national policy needed to be established that would encompass the majority of industries; and (2) states had generally failed to meet their voluntary obligations for health and safety in the workplace.

To solve these shortcomings, the Occupational Safety and Health Act (OSH Act) (designed to "assure so far as possible every working man and woman in the Nation safety and healthful working conditions and to preserve our human resources") was signed by President Nixon on December 29, 1970. Since its effective date on April 28, 1971, this single act has had an enormous impact on the safety and health movement within the United States, more than any other legislation. The law affects approximately 60 million employees in over four million establishments, but excludes employees of state and federal government, who are protected under regulations similar to those within OSHA.

IMPORTANT POINT

It is common, but incorrect, to refer to the OSH Act with the term OSHA. OSHA stands for Occupational Safety and Health Administration, and OSH Act for the Occupational Safety and Health Act. The two acronyms have different meanings.

Under the secretary of labor (a presidential appointment), the assistant secretary of labor for occupational safety and health has the responsibility to guide and administer the Occupational Safety and Health Administration (OSHA).

Under the provisions of the act, each *employer* covered by the act has certain responsibilities. The following list is a summary of the most important ones.

- Provide a workplace free from serious recognized hazards and comply with standards, rules, and regulations issued under the OSH Act. Specifically, the employer has a "general duty" to furnish each employee employment and places of employment that are free from recognized hazards that cause or are likely to cause death or serious physical harm. (This means that even if a hazard in the workplace is not specifically covered by a regulation, the employer must protect the employee anyway). This is commonly known as the Act's *general duty clause*. Safety professionals, and in particular OSHA professionals, view it as a "safety net" (in practice, we have overheard supervisors who are not in compliance with any safety rules, regulations, and/or practices refer to the general duty clause as OSHA's "gotcha" rule). Simply, the employer has the specific duty of complying with safety and health standards promulgated under the act.

- Examine workplace conditions to make sure they conform to applicable OSHA standards.
- Make sure employees have and use safe tools and equipment and properly maintain this equipment.
- Use color codes, posters, labels, or signs to warn employees of potential hazards.
- Establish or update operating procedures and communicate them so that employees follow safety and health requirements.
- Provide medical examinations and training when required by OSHA standards.
- Post, at a prominent location within the workplace, the OSHA poster (or the state-plan equivalent) informing employees of their rights and responsibilities.
- Report to the nearest OSHA office within eight hours any fatal accident or one that results in the hospitalization of three or more employees.
- Keep records of work-related injuries and illnesses. (Note: Employers with 10 or fewer employees and employers in certain low-hazard industries are exempt from this requirement.)
- Provide employees, former employees, and their representatives access to the Log of Work-Related Injuries and Illnesses (OSHA Form 300).
- Provide access to employee medical and exposure records to employees or their authorized representatives.
- Provide to the OSHA compliance officer the names of authorized employee representatives who may be asked to accompany the compliance officer during an inspection.
- Not discriminate against employees who exercise their rights under the act.
- Post OSHA citations at or near the work area involved. Each citation must remain posted until the violation has been corrected, or for three working days, whichever is longer. Post abatement verification documents or tags.
- Correct cited violations by the deadline set in the OSHA citation and submit required abatement verification documentation.

Under the provisions of the act, each *employee* has the duty to comply with the safety and health standards, and with all rules, regulations, and orders applicable to his own actions and conduct on the job. (Experience shows us that when employees are informed of this requirement under the OSH Act, they are usually surprised; they often view the act as only applying to the employer.) An employee should do the following:

- Read the OSHA poster at the job site.
- Comply with all applicable OSHA standards.

## OSHA WORKPLACE POSTER—PUBLICATION 3165

The OSHA poster, also known as the OSHA notice of employee rights, is required to be displayed in every workplace in America.

- Follow all lawful employer safety and health rules and regulations, and wear or use prescribed protective equipment while working.
- Report hazardous conditions to the supervisor.
- Report any job-related injury or illness to the employer, and seek treatment promptly.
- Exercise rights under the act in a responsible manner.

### OSHA REGULATIONS/STANDARDS

OSHA regulations/standards (compliance with which is the major concern of this text) take two basic forms: they are either *specific standards* or *performance standards*. Specific standards explain exactly how to comply. For example, the OSHA regulation covering means of egress from buildings very specifically lists requirements for means of egress, exit access, exit discharge, and so forth. A performance standard lists the ultimate goal of compliance, but does not explain exactly how to accomplish it. A good example of a performance standard is the general duty clause, which states that the employer must protect the health and safety of the employee, even if no OSHA regulation currently covers the work activity in question. These standards do not explain how to accomplish this—that is left up to the employer.

### KEY POINT

We stress compliance with OSHA standards throughout this text because, ultimately, the responsible person in charge of ensuring worker safety must also ensure compliance with these standards. We describe in detail exactly what these compliance requirements are.

## WHAT IS A STATE OSHA PROGRAM?

According to the U.S. Department of Labor (DOL), Section 18 of the OSH Act of 1970 encourages states to develop and operate their own job safety and health programs. OSHA approves and monitors state plans and provides up to 50 percent of an approved plan's operating costs.

There are currently 22 states and jurisdictions operating complete state plans (covering both the private sector and state and local government employees) and 4 (shown below with asterisks)—Connecticut, New Jersey, New York, and the Virgin Islands—that cover public sector employees only. (Eight other states were approved at one time but subsequently withdrew their programs.) These 26 states are as follows:

| | |
|---|---|
| Alaska | New Mexico |
| Arizona | New York* |
| California | North Carolina |
| Connecticut* | Oregon |
| Hawaii | Puerto Rico |
| Indiana | South Carolina |
| Iowa | Tennessee |
| Kentucky | Utah |
| Maryland | Vermont |
| Michigan | Virgin Islands* |
| Minnesota | Virginia |
| Nevada | Washington |
| New Jersey* | Wyoming |

States must set job safety and health standards that are "at least as effective as" comparable federal standards. (Most states adopt standards identical to federal ones.) States have the option to promulgate standards covering hazards not addressed by federal standards.

A state must conduct inspections to enforce its standards, cover public (state and local government) employees, and operate occupational safety and health training and education programs. In addition, most states provide free on-site consultation to help employers identify and correct workplace hazards. Such consultation may be provided either under the plan or through a special agreement under section 21(d) of the act.

## TITLE 29 CFR

Regulations governing labor practices, including safety and health, are listed under Title 29 of the Code of Federal Regulations, with the occupational safety and health regulations found in parts 1900–1999. 29 CFR Part 1910 (General Industry Standards),

Part 1926 (Construction Standards), and Part 1928 (Agriculture) contain the workplace regulations we are concerned with in this text because they are applicable, in one way or another, to the food manufacturing industry.

You do not need to read all the OSHA 29 CFR standards, but you will need to read particular standards at some point. Many OSHA standards are lengthy and complex and some are dated. (Many standards were promulgated in 1970–1972, and in some cases have not been updated since. The standard updating process is ongoing but can be slow and tedious.) Based on personal experience, we feel much of the difficulty in reading the standards is complicated by the use of small type and the failure to separate subsections by some readily identifiable method or to use a reader-friendly format. Quite frankly, another problem (which is debatable, depending on your point of view) is that many of the OSHA standards are vague and ambiguous. That is, many written standards are open to various interpretations—with OSHA's interpretation being the only one that counts. However, all OSHA findings can be appealed through a Notice of Contest (NOC).

Individual parts of Title 29 are designated by a four-digit number (for example, 1928—Agriculture). That "part" designation is the first four numbers of all OSHA standards. A period follows the first four digits. Following that period, each standard is listed in numerical order beginning with number 1. Thus 29 CFR 1928.1 refers to Title 29, Part 1928, Section 1.

The number following the period is the designation given to a particular OSHA standard. In many cases, it will be followed by a name—for example, §1910.146 Permit-required confined spaces. There will then follow various subsections of the standard. They are designated by letters and numbers, all of which are in parentheses. The first subsection is (a)—for example, §1910.146(a). Other major subsections will be similarly designated in alphabetical order—for example, §1910.146(b), §1910.146(c), and so on. However, each of them often has its own subsections. And those subsections, in turn, have subsections.

## DID YOU KNOW?

The format and sequence of all OSHA standards is the same. If you follow the preceding steps, you should be able to locate the particular subsection of the standard you are looking for.

OSHA standards also include subparts. So that you begin to understand what is contained in one of these parts, let's take a look at Part 1910, which is divided into subparts A to Z, shown in table 2.1.

**Table 2.1.   Outline of 19 CFR 1910, Subparts A–Z**

| | |
|---|---|
| Subpart A | General—includes the provisions for OSHA's initial implementation of regulations. |
| Subpart B | Adoption and Extension of Established Federal Standards—explains which businesses are covered by OSHA regulations. |
| Subpart C | General Safety and Health Provisions—provides the right for an employee to gain access to exposure and medical records. |
| Subpart D | Walking and Working Surfaces—establishes requirements for fixed and portable ladders, scaffolding, manually propelled ladder stands, and general walking surfaces. |
| Subpart E | Means of Egress—establishes general requirements for employee emergency plans and fire prevention plans. |
| Subpart F | Powered Platforms, Man Lifts, and Vehicle-Mounted Work Platforms—mandates the minimum requirements for an elevated safe work platform. |
| Subpart G | Occupational Health and Environmental Control—mandates engineering controls of physical hazards such as ventilation for dusts, control of noise, and control of ionizing and nonionizing radiation. |
| Subpart H | Hazardous Materials—provides requirements for the use, handling, and storage of hazardous materials. |
| Subpart I | Personal Protective Equipment—provides general requirements for personal protective equipment. |
| Subpart J | General Environmental Controls—mandates the requirements for sanitation, accident prevention signs and tags, confined space entry, lockout/tagout requirements for hazardous energy. |
| Subpart K | Medical and First Aid—requires that an employer provide first-aid facilities or personnel trained in first aid to be at the facility. |
| Subpart L | Fire Protection—mandates portable or fixed fire suppression systems for workplaces. |
| Subpart M | Compressed Gas and Compressed Air Equipment—presents the requirements for air receivers. |
| Subpart N | Materials Handling and Storage—covers the uses of mechanical lifting devices, changing a flat tire, forklift and helicopter operation. |
| Subpart O | Machinery and Machine Guarding—provides requirements for guarding rotating machinery. |
| Subpart P | Hand and Portable Powered Tools and Other Hand-Held Equipment. |
| Subpart Q | Welding, Cutting, and Brazing—requires the use of eye protection, face shields with arc lenses, proper handling of oxygen and acetylene tanks. |
| Subpart R | Special Industries—covers special requirements for Textiles, Bakery Equipment, Laundry Machinery, Sawmills, Pulpwood Logging, Grain Handling, and Telecommunications. |
| Subpart S | Electrical—requires the use of protection mechanisms for electrical installations. |
| Subpart T | Commercial Diving Operation—mandates requirements for the dive team. |
| Subpart U | Not currently assigned. |
| Subpart Z | Toxic and Hazardous Substances—requires monitoring and protective methods. for controlling hazardous airborne contaminants. |

OSHA uses various provisions in the OSH Act to, for example, determine how well your program works at your food manufacturing site (or any other covered workplace) by reviewing your company's injury data and insurance costs (if your program is effective, your injury and insurance costs will reflect that). Obviously, to get accurate data upon which to base its judgment, OSHA requires extensive record keeping of written programs, injuries, illnesses, safety audits, inspections, corrections, and training. OSHA's record-keeping requirements are discussed in detail later in the text, but for now, let's take a quick look at training.

Training is a major part of the OSH Act. Almost every regulation requires some sort of transmission of information and training. Why? Simply because injury statistics show that newer employees without adequate training are far more likely to be injured on the job than those with more experience and training. Think about it. How well will a brand-new employee operate a complex piece of food manufacturing machinery for the first time? How safe will this same worker be without proper training? OSHA takes a close look at the employer's training records to ensure training is being accomplished. OSHA also observes employees at work to determine if they are performing work tasks in a safe manner. Also, OSHA selects various workers to interview; the OSHA auditor asks the workers questions designed to measure workers' level of training.

Speaking of OSHA auditors (by the way, they like to be called compliance officers), they are the personnel responsible for the enforcement of the provisions of the OSH Act. They enforce the standards by conducting inspections (audits), during which they may award citations and levy civil penalties. These three increasingly punitive steps are designed to achieve a safe workplace by requiring the removal of hazards—requiring employers to comply with the requirements of the act. If hazardous situations are discovered, OSHA often conducts follow-up inspections to assure that the appropriate corrections have been made. OSHA investigates and writes citations based on inspections of the work site.

An OSHA inspector may visit a site based on the following:

- an employee complaint
- a report that an injury or fatality has occurred
- a random visit to a high-risk business

If the inspection uncovers one or more violations, the OSHA compliance officer provides an explanation on a written inspection report. The types of violations include the following:

- **de minimis.** A condition that has no direct or immediate relationship to job safety and health (for example, an error in interpretation of a regulation)
- **General.** Inadequate or nonexistent written programs, lack of training, training records, etc.
- **Repeated.** Violations in which, upon reinspection, another violation is found of a previously cited section of a standard, rule, order, or condition violating the general duty clause
- **Serious.** A violation that could cause serious harm or permanent injury to the employee, and in which the employer did not know, or could not have known, of the violation

- **Willful.** A violation in which evidence shows that the employer knew that a hazardous condition existed that violated an OSHA regulation, but made no reasonable effort to eliminate it
- **Imminent danger.** A condition in which there is reasonable certainty that an existent hazard can be expected to cause death or serious physical harm immediately, or before the hazard can be eliminated through regular procedures

When a compliance officer believes an employer has violated a safety or health requirement of the act, or any standard, rule, or order promulgated under it, the officer will issue a citation. Any citation issued for noncompliance must be posted in clear view near the place where the violation occurred for three working days or until corrected, whichever is longer.

Does the employer have any recourse when cited by OSHA? Actually, the employer can take either of the following courses of action regarding citations:

1. The employer can agree with the citation and correct the problem by the date given on the citation and pay any fines; or
2. The employer can contest the citation, proposed penalty, or correction date, as long as it is done within 15 days of the date the citation for the matter in question is issued.

## THE BOTTOM LINE ON OSHA COMPLIANCE

The disciplines that make up safety, health, and environmental control have become very sophisticated and highly technical, primarily because of the passage of the OSH Act. As a result, highly educated and skilled safety professionals are needed to administer and direct these programs, to maintain a strong position in all required areas of compliance. Experience shows that if a fully qualified individual fills a company safety engineer's position, and if that person has the authority and the budget to manage the program, OSHA is more commonly viewed as a resource—not as an enforcer.

It is important to keep in mind that OSHA's regulatory requirements came about as the solution to disturbing levels of loss of health and life from industrial accidents and dangerous working conditions. Leaving worker safety solely in the hands of employers did not work then, and lest we forget that, assuming that "in this day and age" we are too enlightened to need regulation to keep industry safety standards high, remember that financial concerns are the ultimate bottom line, and improving that bottom line is business's bottom line. Safety is expensive. The way to keep safety of value for industry is to ensure that unsafe conditions are even more costly.

## REFERENCES AND RECOMMENDED READING

*Analysis of Workers' Compensation Laws.* Annual. Washington, D.C.: Chamber of Commerce of the United States.

Brauer, R. L. 1994. *Safety and Health for Engineers.* New York: Van Nostrand Reinhold.

CoVan, J. 1995. *Safety Engineering.* New York: Wiley.

Ferry, T. 1990. *Safety and Health Management Planning.* New York: Van Nostrand Reinhold.

Occupational Safety and Health Administration (OSHA). 2005. OSHA Fines Frankfort, N.Y., Food Plant for Confined Space Hazards: $142,200 in Fines Proposed against Granny's Kitchens, October 3. www.osha.gov/pls/oshaweb/owadisp.show_document?p_table= NEWS_RELEASES&p_id=11613 (accessed July 27, 2007).

*Right Off the Docket.* 1986. Cleveland, OH: Penton.

Spellman, F. R. 1998. *Surviving an OSHA Audit: A Manager's Guide.* Lancaster, PA: Technomic.

# Hazards in the Meatpacking Industry

## NATURE OF THE MEATPACKING INDUSTRY

According to the U.S. Department of Labor's Safety and Health Guide for the Meatpacking Industry (1988):

The meatpacking industry, which employs over a million workers, is considered to be one of the most hazardous industries in the United States. According to the Bureau of Labor Statistics (BLS) this industry had the highest injury rate of any industry in the country for five consecutive years (1980–1985), with a rate three times that of other manufacturing industries.

BLS studies also showed that for 1985, 319 workers were injured during the first month of employment in the industry. Of those, 29 percent were cut by knives or machinery and 30 percent received sprains and strains. In addition, more than 30 percent of all injuries occurred to workers 25 yeas of age or younger. Younger, new workers are at the highest occupational risk and suffer a significant proportion of all injuries.

Workers can be seriously injured by moving animals prior to stunning, and by stunning guns that may prematurely or inadvertently discharge while they try to still the animal. During the hoisting operation, it is possible for a 2,000-pound carcass to fall on workers and injure them if faulty chains break or slip off the carcass's hind leg. Workers can suffer from crippling arm, hand, and wrist injuries. For example, carpal tunnel syndrome, caused by repetitive motion, can literally wear out the nerves running through one section of the wrist. Workers can be cut by their own knives and by other workers' knives during the butchering process. Back injuries can result from loading and unloading meat from trucks and from moving meat, meat racks, or meat trees along overhead rails. Workers can be severely burned by cleaning solvents and burned by heat sealant machines when they wrap meat. It is not uncommon for workers to sever fingers or hands on machines that are improperly locked-out or inadequately guarded. For example, in 1985, BLS studies reported 1,748 cases of injuries to the fingers, including 76 amputations. Workers can also injure themselves by falling on treacherously slippery floors and may be exposed to extremes of heat and cold.

This publication is designed to increase employer and employee awareness of these and other workplace hazards and to highlight the ways in which employer and employees can work together to eliminate them. Employers are encouraged to review and strengthen overall safety and health precautions to guard against workplace accidents, injuries, and illnesses.

## POTENTIAL HAZARDS

Machinery such as head splitters, bone splitters, snout pullers, and jaw pullers, as well as band saws and cleavers, pose potential hazards to workers during the various stages of processing animal carcasses. A wide variety of other occupational safety and health hazards exists in the industry (see table 3.1). These hazards are identified and discussed in the following paragraphs.

### Knife Cuts

Knives are the major causes of cuts and abrasions to the hands and the torso. Although modern technology has eliminated a number of hand knife operations, the

## KNIFE ACCIDENTS

- A worker used a knife to pick up a ham prior to boning; the knife slipped out of the ham, striking him in the eye and blinding him (NSC 1979, 33).
- Another worker was permanently disfigured when his knife slipped out of a piece of meat and struck his nose, upper lip, and chin (NSC 1979, 33).
- Workers have also been cut by other workers as they remove their knives from a slab of meat. These "neighbor cuts" are usually the direct result of overcrowded working conditions (OSHA 1988).

hand knife remains the most commonly used tool and causes the most frequent and severe accidents (OSHA 1988).

### Falls

Falls also represent one of the greatest sources of serious injuries. Because of the nature of the work, floor surfaces throughout the plants tend to be wet and slippery. Animal fat, when allowed to accumulate on floors to dangerous levels, and blood, leaking pipes, and poor drainage are the major contributors to treacherously slippery floors.

### Back Injuries

Back injuries tend to be more common among workers in the shipping department. These employees, called "luggers," are required to lug or carry on their shoulders carcasses (weighing up to 300 pounds) to trucks or railcars for shipment.

### Toxic Substances

Workers are often exposed to ammonia. Ammonia is a gas with a characteristic pungent odor and is used as a refrigerant, and occasionally, as a cleaning compound. Leaks can occur in the refrigeration pipes carrying ammonia to coolers. Contact with anhydrous liquid ammonia or with aqueous solution is intensely irritating to the mucous membranes, eyes, and skin. There may be corrosive burns to the skin or blister formation. Ammonia gas is also irritating to the eyes and to moist skin. Mild to moderate exposure to the gas can produce headaches, salivation, burning of the throat, perspiration, nausea, and vomiting. Irritation from ammonia gas to the eyes and nose may be sufficiently intense to compel workers to leave the area. If escape is not possible, there may be severe irritation of the respiratory tract with the production of cough, pulmonary edema, or respiratory arrest. Bronchitis or pneumonia may follow a severe exposure.

On some occasions, employees have been exposed to unsafe levels of carbon dioxide from the dry ice used in the packaging process. When meat is ready to be frozen for packaging, it is put into vats where dry ice is stored. During this process, carbon dioxide gas may escape from these vats and spread throughout the room. Breathing high levels of this gas causes headaches, dizziness, nausea, vomiting, and even death.

Workers are also exposed to carbon monoxide. Carbon monoxide is a colorless, odorless gas that is undetectable by the unaided senses and is often mixed with other gases. Workers are exposed to this gas when smokehouses are improperly ventilated. Overexposed workers may experience headaches, dizziness, drowsiness, nausea, vomiting, and death. Carbon monoxide also aggravates other conditions, particularly heart disease and respiratory problems.

Workers are also exposed to the thermal degradation products of polyvinyl chloride (PVC) food-wrap film. PVC film used for wrapping meat is cut on a hot wire, wrapped around the package of meat, and sealed by the use of a heated pad. When the PVC film is heated, thermal degradation products irritate workers' eyes, nose, and throat or cause more serious problems such as wheezing, chest pains, coughing, difficulty in breathing, nausea, muscle pains, chills, and fever.

### Cumulative Trauma Disorders

Cumulative trauma disorders are widespread among workers in the meatpacking industry. Cumulative trauma disorders such as tendonitis (inflammation of a tendon sheath) and carpal tunnel syndrome are very serious diseases that often afflict workers whose jobs require repetitive hand movement and exertion.

Carpal tunnel syndrome is the disorder most commonly reported for this industry and is caused by repeated bending of the wrist combined with gripping, squeezing, and twisting motions. A swelling in the wrist joint causes pressure on a nerve in the wrist. Early symptoms of the disease are tingling sensations in the thumbs and in the index and middle fingers. Experience has shown that if workers ignore these symptoms, sometimes misdiagnosed as arthritis, they could experience permanent weakness and numbness in the hand coupled with severe pain in the hands, elbows, and shoulders.

### Infectious Diseases

Workers are also susceptible to infectious disease such as brucellosis, erysipeloid, leptospirosis, dermatophytosis, and warts. Brucellosis is caused by a bacterium and is transmitted by the handling of cattle or swine. Persons who suffer from this bacterium experience constant or recurring fever, headaches, weakness, joint pain, night sweats, and loss of appetite.

Erysipeioia and leptospirosis are also caused by bacteria. Erysipeloid is transmitted by infection of skin puncture wounds, scratches and abrasions; it causes redness and irritation around the site of infection and can spread to the bloodstream and lymph nodes. Leptospirosis is transmitted through direct contact with infected animals or through water, moist soil, or vegetation contaminated by the urine of infected animals. Muscular aches, eye infections, fever, vomiting, chills, and headaches occur, and kidney and liver damage may develop.

Dermatophytosis, on the other hand, is a fungal disease and is transmitted by contact with the hair and skin of infected persons and animals. Dermatophytosis, also know as ringworm, causes the hair to fall out and small yellowish cuplike crusts to develop on the scalp (OSHA 1988).

### DID YOU KNOW?

*Verruca vulgaris*, a wart caused by a virus, can be spread by infectious workers who have contaminated towels, meat, fish knives, worktables, or other objects. (OSHA 1988)

## THE MEATPACKING INDUSTRY AND OSHA

There are currently no specific OSHA standards for the meatpacking industry. However, in the following we highlight those general industry standards related to and/or applicable to the meatpacking industry.

### General Industry (29 CFR 1910)
- 1910 Subpart D, Walking-working surfaces
  - 1910.21, Definitions
  - 1910.22, General requirements
  - 1910.23, Guarding floor and wall openings and holes
  - 1910.24, Fixed industrial stairs
  - 1910.25, Portable wood ladders
  - 1910.26, Portable metal ladders
  - 1910.27, Fixed ladders
- 1910 Subpart G, Occupational health and environmental control
  - 1910.95, Occupational noise exposure

- 1910 Subpart H, Hazardous materials
  - 1910.119, Process safety management of highly hazardous chemicals
- 1910 Subpart I, Personal protective equipment
- 1910 Subpart J, General environmental controls
  - 1910.147, The control of hazardous energy (lockout/tagout)
- 1910 Subpart O, Machinery and machine guarding
  - 1919.212, General requirements for all machines
    - 1910.212(a)(3)(ii)
- 1910 Subpart P, Hand and portable powered tools and other hand-held equipment
  - 1910.243, Guarding of portable powered tools
- 1910 Subpart S, Electrical
- 1910 Subpart Z, Toxic and hazardous substances
  - 1910.1200, Hazard communication

## THE JUNGLE REVISITED II

Upton Sinclair's *The Jungle* exposed real-life conditions in meatpacking plants to a horrified public. But what most shocked the popular conscience was Sinclair's portrayal of vermin, animal feces, human blood and body parts going into meat people ate, and the deceptive practices used to sell such adulterated products. (Young 1989)

### EMPLOYER AND EMPLOYEE RESPONSIBILITIES

An employer's commitment to a safe and healthful environment is essential in the reduction of workplace injury and illness. This commitment can be demonstrated through personal concern for employee safety and health, by the priority placed on safety and health issues, and by setting good examples for workplace safety and health. Employers should also take any necessary corrective action after an inspection or accident. They should assure that appropriate channels of communication exist to allow information and feedback on safety and health concerns and performance. In addition, regular self-inspections of the workplace will further help prevent hazards by assuring that established safe work practices are being followed and that unsafe conditions or procedures are identified and corrected properly. These inspections are in addition to the everyday safety and health checks that are part of the routine duties of supervisors.

Since workers are also accountable for safety and health, it is extremely important that they too have a strong commitment to workplace safety and health. Workers should immediately inform their supervisor or their employer of any hazards that exist in the workplace and of conditions, equipment, and procedures that would be potentially hazardous. Workers should also understand what the safety and health program is all about, why it is important to them, and how it affects their work.

Finally, employers who want help in recognizing safety and health hazards and in improving their safety and health programs can receive assistance from a free consultation service largely funded by the Occupational Safety and Health Administration. The service is delivered by state governments using well-trained professional staff. The service offers advice and help in correcting problems and in maintaining continued effective protection. In addition to helping employers identify and correct specific hazards, consultants provide guidance in establishing or improving an effective safety and health program and offer training and education for the company, the supervisors, and the employees. Such consultation is a cooperative approach to solving safety and health problems in the workplace. As a voluntary activity, it is neither automatic nor expected; it must be requested (OSHA 1988).

## HAZARD CONTROL METHODS

The unique safety and health hazards found in this industry can be minimized or eliminated with the proper use of control methods. A preferred way of controlling potential hazards is through the use of *engineering controls*. Engineering controls are methods that prevent harmful worker exposure through proper design of equipment and processes.

### Guardrails

The use of guardrails can protect workers from accidental falls. Open surface dip tanks, used for sterilizing shackling equipment, and elevated work platforms must have guardrails. Railings should also be checked to see that they are securely attached to walls.

### Floors

Employers should install a nonskid flooring material or rubberized cushioned floor mats at all workstations for workers to stand on, especially in areas where hand knives and power tools are used.

### Wiring

All electrical wiring should be checked periodically for cracking, fraying, or other defects, and all electrical equipment should be grounded (OSHA 1988).

All electrical wiring must be in compliance with the requirements of the National Fire Protection Association Standard for Electrical Protection in the Workplace.

- **NFPA 70E Article 215**
  - 215.1—Covers for wiring system components shall be in place with all associated hardware, and there shall be no unprotected openings.
  - 215.2—Open wiring protection, such as location or barriers, shall be maintained to prevent accidental contact.
  - 215.3—Raceways and cable trays shall be maintained to provide physical protection and support for conductors.

### Equipment and Machine Guarding

Equipment used to hold and move meat and items such as shackles, conveyors, and hooks should be checked frequently and repaired. Equipment that poses a hazardous energy source should, when not in use, be subject to lock-out and tag-out procedures. This assures that workers inspecting or maintaining equipment are not injured by start-up of the equipment. All equipment that poses a hazard should be guarded.

### Local Exhaust Ventilation

A preferred control method for removing air contaminants from the workplace is local exhaust ventilation. This control is located at the source of the generation of contaminants, and captures, rather than dilutes, the hazardous substances before they escape into the workplace environment.

### General Ventilation

General, or dilution, ventilation systems are also recommended because they add or remove air from the workplace to keep the concentration of air contaminants below hazardous levels. General ventilation consists of air flow through open windows or doors, fans, and roof ventilators. General ventilation control only dilutes air contaminants, unlike local exhaust ventilation, which removes air contaminants. When using general ventilation systems, care should be taken not to recirculate the air contaminants throughout the workplace.

### Administrative Controls

An employer also might decide to use administrative controls to minimize the risk of carpal tunnel syndrome, back and shoulder injuries, and exposure to toxic substances. One type of administrative control would be to reduce employee work periods in which excessive repetitive wrist bending is necessary or when the worker is exposed to hazardous substances.

### Work Practices

Safe work practices are essential in helping to maintain a safe and healthful work environment. Workers must therefore be encouraged and be given sufficient time and equipment to keep surfaces clean and orderly.

To do this, spills must be cleaned up immediately. Water, blood, or grease on floors will cause falls. Also, wet working conditions pose a serious threat of electrocution. Periods during the day should also be set aside for general housekeeping, and constant surveillance should be kept to spot slippery areas. Nonskid floor mats can also be used successfully in potentially dangerous areas. Knives left carelessly in sinks or on counters can cause serious accidents. Knives should be kept sharpened at all times. Dull knives can cause serious safety hazards and worker fatigue. Equipment such as the band saw and the bacon press must be cleaned with the power off and locked-out, and tagged-out. Workers should use only tools and equipment with which they are familiar. Moreover, employers should check refrigeration systems regularly for leaks and should make sure that hazardous substances, such as ammonia, are identified by appropriate hazard warnings (labels, signs, etc.).

Employers should make handwashing facilities readily available to employees working with or near toxic substances. It is equally important that handwashing facilities be made available for workers who handle meat without the use of protective gloves. Prompt handwashing and the use of disposable hand towels will help prevent the spread of infectious diseases.

### Protective Clothing and Equipment

Since slippery floors are a major cause of falls, protective clothing such as safety shoes or boots with toe guards and slip-resistant soles must be worn by workers. To help reduce the spread of infectious diseases, protective gloves should be worn when workers handle meat. Workers who use cleaning compounds must also wear protective gloves to prevent chemical burns. In addition, workers who use knives must be provided with metal mesh gloves and aprons, and wrists and forearm guards to protect them from knife cuts.

Workers performing hoisting and shackling operations should be protected with safety helmets that meet the specifications of American National Standard Requirements for Protective Head Wear for Industrial Workers, as well as a barricade or shield assembly. These safeguards can prevent injury from falling or moving animals and/or materials. In addition, removing the worker from the immediate area during hoisting operations is recommended.

The employer must furnish employees with proper personal protective equipment required for the specific work operation and exposure. For example, in the event of exposure to toxic chemicals, a worker must be provided with a suitable respirator to prevent inhalation of harmful substances.

In addition, adjustable work stands should be made available to accommodate for worker height to minimize the possibility of back strain.

Machines and equipment found in meatpacking plants produce high levels of noise; in such circumstances, workers must be provided with ear plugs. The employer may be required to provide workers with face shields or goggles when workers mix or handle cleaners. The use of this equipment will prevent chemical burns to the face and eyes. Goggles may also be required during the boning, trimming, and cutting operations to prevent foreign objects from entering the workers' eyes (OSHA 1988).

Many of the operational hazards in the meatpacking industry are listed in table 3.1.

## ADMINISTRATIVE REQUIREMENTS/PRACTICES

OSHA mandates a number of administrative requirements for the meatpacking industry (as is does for most major industries), including the need to provide training and maintain various records. In this section we describe these administrative require-

**Table 3.1. Operational hazards in the meatpacking industry**

| Operation Performed | Equipment/Substances | Accidents/Injuries |
|---|---|---|
| Stunning | Knocking gun | Severe shock, body punctures |
| Skinning/Removing front legs | Pincher device | Amputations, eye injuries, cuts, falls |
| Splitting animal | Splitter saws | Eye injuries, carpal tunnel syndrome, amputations, cuts, falls |
| Removing brain | Head splitter | Cuts, amputations, eye injuries, falls |
| Transporting products | Screw conveyors, screw auger | Fractures, cuts, amputations, falls |
| Cutting/trimming/boning | Hand knives, saws—circular saw, band saw | Cuts, eye injuries, carpal tunnel syndrome, falls |
| Removing jaw bone/snout | Jaw bone, snout puller | Amputations, falls |
| Preparing bacon for slicing | Bacon/belly press | Amputations, falls |
| Tenderizing | Electrical meat tenderizers | Severe shock, amputations, cuts, eye injuries |
| Cleaning equipment | Lock-out, tag-out | Amputations, cuts |
| Hoisting/shackling | Chain/dolly assembly | Falls, falling carcasses |
| Wrapping meat | Sealant machine/polyvinyl chloride, meat | Exposure to toxic substances; severe burns to hands/arms, falls |
| Lugging meat | Carcasses | Severe back/shoulder injuries, falls |
| Refrigeration/curing, cleaning, wrapping | Ammonia, carbon dioxide, carbon monoxide, polyvinyl chloride | Upper respiratory irritation and damage |

Source: Compiled by the U.S. Department Occupational and Health Administration, based on information from "The Unique Hazards of Packing," by Jeff Spahn, area director, U.S. Department of Labor, Occupational Safety and Health Administration, Wichita, Kansas, 1976 (unpublished information; OSHA 1988).

ments. Before doing this, however, it is important to point out that there is a definite need to indoctrinate new meatpacking employees to the hazards they may face before putting them to work—before placing them in harm's way. We have found that a well-designed and properly presented new-employee orientation program that stresses the need for safety can be extremely beneficial.

### Training and Education

BLS studies show that many workers are injured because they often do not receive the safety training they need, even on jobs involving dangerous equipment (e.g., the meatpacking industry) where training is clearly essential. These studies also show that younger, and especially new, employees are most at risk because they are not taught the necessary safety measures before they begin work. (Again, we stress the need for some form of formal new-employee orientation training before assigning new hires to dangerous work.) More experienced workers, on the other hand, are injured because the task becomes routine, and they are not as cautious as they might be otherwise.

It is essential, therefore, that employers develop, implement, and maintain at the workplace a written comprehensive training program for all employees. A comprehensive, well-organized training program helps the employer to educate workers in safe work practices and techniques, and helps demonstrate the employer's concern for, and commitment to, safe work practices (OSHA 1988).

#### DID YOU KNOW?

The Secretary of Labor's Hazardous Occupations Order Number 10 sets a minimum age of 18 years for employment in many meat-processing occupations. See CFR, Part 570.61, December 29, 1971.

The training program should inform workers about safety and health hazards and their prevention, the proper use and maintenance of equipment, any appropriate work practices, a medical surveillance program, and emergency situations.

Once the employer has developed and implemented the safety and health program, he or she should choose a person who is committed to workplace safety and health to manage the program. This individual should have time to devote to developing and managing the program and must be willing to take on the responsibility and the accountability that goes with operating an effective program.

The employer should also make all employees familiar with their surroundings and work environment. Furthermore, employers should train workers annually in their work tasks or in new job assignments that expose them to new hazards. New employees should be trained at the time of initial assignment, and annually thereafter (OSHA 1988).

## IMPORTANT POINT

More training is needed when new equipment, materials, or processes are introduced, when procedures are updated or revised, or when employee performance is inadequate. (OSHA 1988)

### First-Aid Training

At least one person on each shift should be trained and certified in first aid. First-aid training should include, as a minimum, completion of an approved first-aid training course. Moreover, proper first-aid supplies must be readily available for emergency use. Prearranged ambulance services should also be available for any emergency.

### Hazard Communication (29 CFR 1910.1200)

OSHA's Hazard Communication Standard requires the employer to establish a written hazard communication program to transmit information on the chemical hazards to which employees are exposed. The program must include container labeling and other forms of warning and must provide exposed workers with material safety data sheets (MSDSs). The MSDS details the substance's properties and the nature of the hazard. The standard further requires chemical hazard training for exposed workers.

In addition, workers who handle only sealed containers of chemicals are required to keep labels affixed to incoming containers, are to be provided access to MSDSs while the material is in the workplace, and are to be trained on how to protect themselves against hazards (OSHA 1988).

## DID YOU KNOW?

MSDSs must be made available to employees/outside contractors/site visitors 24 hours/day.

### Access to Exposure and Medical Records

In 1988, OSHA issued a standard that requires employers to provide employees with information to assist in the management of their own safety and health. The standard, "Access to Employee Exposure and Medical Records" (29 CFR 1910.20), permits direct access by employees or their designated representatives and by OSHA to employer-maintained exposure and medical records. This access is designed to yield both direct and indirect improvements in the detection, treatment, and prevention of occupational disease. Access to these records should result in a decreased incidence of occupational exposure and should aid in designing and implementing new control measures.

### Record Keeping

OSHA requires employers with 11 or more employees to prepare and maintain pertinent injury and illness records of accidents affecting their employees. Moreover, all employers are required to report to the nearest OSHA office, within [eight] hours, all accidents resulting in a work-related death or in [three] or more hospitalizations. The report may be either oral or written.

The employer is also required to maintain occupational injury and illness records at each workplace. Records must be retained for five calendar years following the end of the year to which they relate and may be inspected and copied at any reasonable time by authorized federal or state government representatives.

These records are important to the employer in analyzing the effectiveness of safety and health programs. They are also important to OSHA inspectors in deciding whether to conduct a complete workplace inspection, and if so, where to concentrate their attention.

### Emergency Response

Proper planning for emergencies is necessary to minimize employee injury. It is important, therefore, that employers in the meatpacking industry develop and implement a written emergency action plan. The plan should include elements such as (1) emergency escape procedures and emergency escape route assignments, (2) procedures for employees to follow who remain to perform (or shut down) critical plant operations prior to evacuation, (3) procedures to account for all employees after emergency evacuation has been completed, (4) assignment of rescue and medical duties to those employees who are to perform them and the preferred means of reporting emergencies, and (5) names and regular job titles of persons or departments to be contacted for further information or explanation of duties under the plan. Emergency phone numbers should also be posted in a conspicuous place near or on telephones (OSHA 1988).

## THE JUNGLE REVISITED III

The American public paid little attention to the treatment of the central character of Sinclair's novel, a Lithuanian immigrant named Jurgis Rudkus who worked in the meatpacking plants. Sinclair reportedly lamented, "I aimed at the public's heart, and I hit it in the stomach" (Rasmussen 2003). Inside the plants, workers faced constant wage cuts, production line speedup, injuries and disease, and instant dismissal and blacklisting if they protested conditions. Outside, they lived with no medical care, no education, and no decent housing. Rudkus and his coworkers were "aliens" both legally and culturally; not citizens, unable to speak good English, ignorant of their rights, and afraid to turn to governmental authorities for help. A century later, abusive working conditions and treatment sometimes still torment a mostly immigrant labor force in the American meatpacking industry.

## REFERENCES AND RECOMMENDED READING

Bureau of Labor Statistics (BLS). 1985. *Supplementary Data System.* Washington, DC: Department of Labor.

National Safety Council (NSC). 1979. *Meat Industry Safety Guidelines.* Chicago: National Safety Council.

Occupational Safety and Health Administration (OSHA). 1988. Safety and Health Guide for the Meatpacking Industry. www.osha.gov/Publications/OSHA3108/osha3108.html (accessed August 2, 2007).

Rasmussen, C. 2003. Muckraker's Own Life as Compelling as His Writing. *Los Angeles Times,* May 11, Metro, p. 4.

Young, J. H. 1989. *Pure Food: Securing the Federal Food and Drugs Act of 1906.* Princeton, NJ: Princeton University Press.

# 4

# Hazards in the Poultry Processing Industry

**Inferno in Hamlet, North Carolina**

In September 1991, 25 people died and 54 were injured as a result of a fire in the Imperial Food Products, Inc., poultry plant in Hamlet, North Carolina. The cause of the fire was the ignition of hydraulic oil from a ruptured line only a few feet from a natural-gas-fired cooker. The cooker was used to cook chicken pieces for distribution to restaurants.

Many OSHA violations were uncovered after the fire. The basic OSHA exit and fire safety violations that contributed to the deaths and injuries were

- locked doors
- no marking of exits or non-exits
- excessive travel distances to exits
- no fire alarms
- obstructed doors
- no emergency action plan or fire prevention plan
- no automatic fire suppression plan

The tragic Hamlet fire received a lot of publicity. In spite of this publicity, blocked exits continue to be found in poultry processing facilities. OSHA cited a plant in Hudson, Missouri, for blocking fire and emergency exits in July 1997. (OSHA 2007a)

## NATURE OF THE POULTRY INDUSTRY

According to Best Food Nation (2007), "poultry is the number one protein purchased by American consumers, at more than 100 pounds per year for every man, woman, and child."

In recent years there has been an increased demand for chicken products. Chicken is relatively lean and has fewer calories and is thus viewed as a healthy alternative to other food types. Moreover, American consumers have demanded the convenience of "fast foods," pre-cut and packaged meats and boneless chicken pieces. The poultry industry (along with others) has had to institute changes to meet these public demands. Changes in the industry have heightened the need for increased safety vigilance and implementation of safe work practices. Appropriate guards around the moving parts of machinery and the blades of saws and knives, adequate ventilation, the use of personal protective equipment, and good house-keeping practices are just a few of the safe practices required.

The poultry industry can include hatcheries and farms where chicks are grown; feed mills where grains are stored, selected, and mixed for hatcheries; and process-ing plants. All fowl (turkey, chicken, duck, capon, quail, etc.) that are processed and made available for consumption are considered to be part of the poultry industry. This text focuses on the processing of chicken, but also applies to the processing of other poultry (NCDOL 2005).

OSHA (2007b) states that the poultry industry can be divided into two stages, "each with its own particular hazards":

The first stage is the raising of live birds to the desired size, with delivery to the processing plant and preparing the live birds for slaughtering. The second stage is the slaughtering, processing, and packaging of the birds. Major hazards in the first stage are generally respiratory hazards resulting from exposure to organic dusts (lit-ter, manure, dander) and ammonia. These are controlled using ventilation and per-sonal protective equipment (PPE). Additional hazards include those associated with agricultural machinery, feed delivery systems, waste removal systems, and er-gonomic hazards especially as birds are prepared for slaughtering. A potentially sig-nificant hazard is the presence of microbiologicals and endotoxins in the organic dusts. In the second stage, common elements in an effective safety and health pro-gram include control of ergonomic hazards (discussed in detail later) to prevent cu-mulative trauma disorders, machine guarding and PPE to prevent cuts, care of walking/working surfaces to reduce trips and falls, design and maintenance of elec-trical systems, and lockout/tagout procedures to prevent accidental startup of ma-chinery (OSHA 2007b).

In regard to OSHA citations—even though there are currently no specific OSHA standards for poultry processing—the most frequently cited General Industry Stan-dards (and guidelines) by federal/state OSHA for the poultry slaughtering and pro-cessing industry include:

- 29 CFR 1910.22—Walking/Working Surfaces—This standard covers floor condi-tions, including wet surfaces.

- Various citations under Section 5(a)(1) of the OSH Act (General Duty Clause)
- Personal Protective Equipment
  - General requirements under Personal Protective Equipment (PPE) standard
  - Sharp knives and utensils in the workplace
- Machine Guarding—29 CFR 1910.212
  - Compliance policy on unguarded rotary knives used in poultry processing industries
- Lockout/Tagout—29 CFR 1910.147
  - Industry full compliance with 29 CFR 1910.147
- 29 CFR Part 1904—Recording and Reporting Occupational Injuries and Illnesses
- 29 CFR 1910.94—Occupational Health and Environment Control—Ventilation, including dust collectors, is addressed by this standard.
- 29 CFR 1910.95—Occupational Noise Exposure
- 29 CFR 1910.132—PPE
- 29 CFR 1910.151 Medical Services and First Aid
- 29 CFR 1910.136—Foot Protection
- 29 CFR 1910.138—Hand Protection
- Stairways and Ladders—OSHA Publication 3124
- 29 CFR 1910 Subpart S, Electrical—NFPA 70(E); National Electrical Code (NEC)
- 29 CFR 1910 Subpart Z—Toxic and Hazardous Substances

## POULTRY PROCESSING PROCEDURES AND HAZARDS

In the OSHA Poultry Processing Industry eTool (2007c), the entire poultry processing procedure is discussed and described in great detail. For the purposes of this text, we have compiled much of this information into a ready, user-friendly format.

As is clearly evident from figure 4, poultry processing is accomplished through a set of step-by-step stages starting with various ongoing *sanitation* tasks (accomplished within/around each procedure) through the *receiving and killing, evisceration, cutting and deboning, packout,* and *warehousing* of product. Each of the steps/stages and the hazards involved with each is discussed briefly here. Possible preventive measures are discussed later.

### DID YOU KNOW?

Since 1975, workers in the poultry processing industry have consistently suffered injuries and illnesses at a rate more than twice the national average.

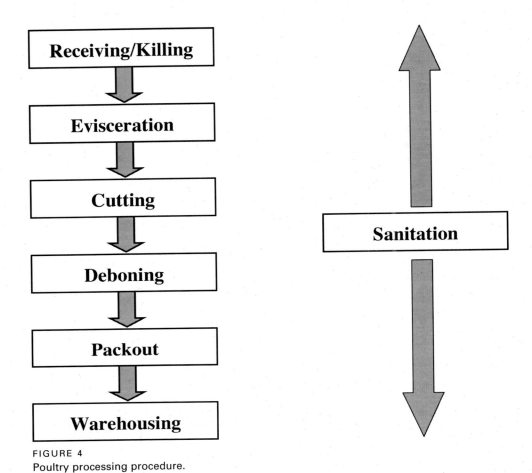

FIGURE 4
Poultry processing procedure.

### Sanitation Worker

The job of the sanitation worker is one of the most hazardous jobs in the poultry processing industry. Sanitation workers may work a regular production shift, or they may be part of a special sanitation or cleaning crew.

The focus of sanitation workers who work a regular production shift is cleaning the machinery and floors. They move product to allow cleaning and use high-pressure water hoses and squeegees to clean the floors. This type of job is frequently an entry-level position. Workers who hold these jobs do not have the experience needed to be familiar with the many hazards of the equipment and the environment in which they work. They need extensive training (OSHA 2007c).

SANITATION WORKER FATALITY INCIDENT #1

A sanitation worker at a poultry plant was cleaning out a chiller. The motor that powers the paddles inside the chiller was accidentally turned on by a coworker who was cleaning a different chiller. Neither the main power control nor the chiller control box had been locked or tagged out in accordance with 29 CFR 1910.147, The Control of Hazardous Energy (Lockout/Tagout). The employee was struck by and caught between the rotating paddle blades and the interior wall of the chiller. He died of severe chest injuries approximately 28 hours after the accident. (OSHA 2007c)

The daily sanitation or clean-up crew has the responsibility of cleaning all product contact surfaces throughout the plant to comply with requirements of the Food Safety and Inspection Service (FSIS), U.S. Department of Agriculture. If the clean-up crew has not done a satisfactory job, the FSIS inspector will not allow the plant to begin a production shift.

When the sanitation crew must remove guards or components to effectively clean processing equipment, and this action (or any other action) exposes crew members to hazardous energy, the equipment must be isolated from its energy source(s), and the energy isolation devices must be locked out or tagged out. In some situations, the equipment must be re-energized for a limited period of time for testing or repositioning purposes. During the testing or positioning period, a sequence of steps must be followed to maintain the integrity of employee protection, and alternative protection, such as removing workers from the machine area, must be provided to prevent employee exposure to machine hazards. Once the testing/positioning activity is completed, the equipment again must be de-energized and locked or tagged out before undertaking further cleaning activities.

Depending on the part of the country in which the plant is located, the sanitation crew may be plant employees or they may work for a contractor. Again, they may be entry-level employees who need extra training to become familiar with the hazards of their work and ways to lessen the hazards.

An additional condition that may contribute to the hazardous nature of the work is that the crew may receive eight hours' pay regardless of how early they finish the job. This gives them an incentive to work as fast as possible and may even contribute to taking shortcuts such as not locking or tagging out equipment.

The sanitation worker is exposed to most of the safety and health hazards throughout the plant, including

- cuts, lacerations, and amputations when removing blades from equipment.
- struck by, struck against, and caught in equipment, such as chiller paddles, or when climbing over or under equipment.
- slips, trips and falls, whether from ladders or climbing on equipment, slipping on wet surfaces, or tripping over drain covers that have been removed and not replaced. Strains, sprains, and/or fractures may result.
- electric shock, which is an increased risk in poultry processing plants because of the wet environment.
- chemical hazards, such as cleaners, that can cause skin or eye irritation or burns (OSHA 2007c).

### SANITATION WORKER FATALITY INCIDENT #2

In preparation for the next day's work, a clean-up crew was using a high-pressure water hose to clean machines in a feather-picking room. One of the two employees was preparing to wash down a feather-picking machine, and the other was washing down a scalder machine. The first employee cranked open his machine to get inside and wash out feathers that had lodged throughout the machine during the feather-picking process. Electric cords were pinched between the metal crank and the metal frame of this machine, and the insulation on the cords was pierced. As a result, an energized conductor was contacting the metal frame of the machine. Because the machine was turned off at the time, the employees were unaware of the condition of the cord. The first employee then moved between the two machines to begin the wash down. When he contacted the energized machine frame, he was electrocuted. His coworker received an electric shock when he tried to pull the first employee from the area. The second employee was hospitalized for his injury. (OSHA 2007c)

### Receiving and Killing

The receiving and killing operation is a largely automated process in most poultry plants and includes receiving live birds, killing, scalding, defeathering, and removing feet. This operation includes the following tasks:

**Task 1: Forklift Operator.** The operator moves the poultry cages from trucks to conveyor/dumping areas. This process moves from the outside to the inside of the dock area. Hazards of this task may include

- **Collision.** Forklifts are driven in heavily congested areas where many other operations are being performed. Visibility and sightlines can be limited, increasing the chance of collision between machinery and employees.
- **Defective forklift.** Over a long period of time the solid rubber wheel on the front of the forklift has developed a flat spot, resulting in unstable loads and poor handling.
- **Slips and falls.** The employee is outside and is exposed to all weather conditions: rain, snow, ice, and heat, resulting in possible slips or falls from working/walking on slippery surfaces.
- **Strains, reaches, and cuts.** Employee must release pull chain to remove chicken cages from trucks, resulting in strains from reaching and possible cuts.

**Task 2a: Automatic Dumper Operator (Automatic Dumper System).** An employee operates the machinery that mechanically dumps live birds from catch-cages. After being removed from cages, the birds are moved by conveyor to the live hang area. Hazards of this task may include

- **Awkward machine controls.** Controls may be difficult to manipulate and/or improperly located. This may cause ergonomic stress on the hands, arms, shoulders, and upper back, and may result in injury.
- **Falls and caught by machinery.** Frequent and rapid movement near heavy machinery increases the chance that employees may slip into the machine or be caught by or between moving parts. This is especially true if machinery is outside or exposed to wet or icy conditions.

**Task 2b: Manual Back Dock Worker.** Some facilities still use smaller catch-cages holding 10 to 12 birds each that are normally emptied manually by the back dock worker. Stacks of cages are moved from the truck to the dock area by forklift. Workers then manually lift and remove cages from stacks and empty poultry out of cages. Birds are transported by conveyor from this station to the live hang area. Hazards of this task may include

- **Lifting, bending, and reaching.** The worker must manually remove cages from stacks, and lift and tilt them to empty chickens from cages. The lifting, bending, and reaching causes stress to the back and shoulders.

- **Disease exposure.** As birds are dumped, feather dander and fecal debris may become airborne and be inhaled by employees. Diseases associated with handling live chickens and contact with poultry feces and dust include allergic alveolitis, cryptosporidiosis, histoplasmosis, hypersensitivity pneumonitis, psittacosis, and Newcastle disease.

## PSITTACOSIS

Psittacosis is caused by a bacterium, Chlamydia psittaci, which is transmitted to humans from birds. "Psittacine" birds, like parrots and parakeets, are classically responsible, although pigeons, chickens, and turkeys may carry the disease as well. An infected bird may appear to have red, watery eyes, nasal discharge, diarrhea, and a poor appetite. After a bird recovers from infection, the bacteria may remain in the blood, feathers, and droppings for many weeks.

Humans may acquire psittacosis by inhaling infected particles from bird droppings. Symptoms begin one to three weeks after exposure, and usually include headache, fever, and cough. A "flu-like" syndrome of nausea/vomiting, joint aches, and muscle aches is also common. Severe infection may develop into pneumonia that requires hospitalization. Psittacosis is treated with common antibiotics (doxycycline or erythromycin), though recovery may take several weeks. Sustained immunity to infection does not develop; some people have been reported to get the disease more than once. Less than 1 percent of all cases are fatal. (OSHA 1994)

**Task 3: Live Hang.** Employees take live birds from conveyors and hang them by their feet from a shackle conveyor. Hazards of this task may include

- **Reaching to conveyor and shackles.** Ergonomic hazards of reaching down to access birds on supply conveyor and reaching up to hang them on the shackle conveyor can lead to shoulder, back, and neck strain because of awkward postures and repetitive motion. The employees at the beginning of the line often work faster than those near the end of the line, causing fatigue. Workers stand for long periods of time.
- **Disease exposure.** Workers get covered with poultry mess and dust that can expose them to diseases associated with handling live chickens and contact with

poultry feces and dust, such as allergic alveolitis, cryptosporidiosis, histoplasmosis, hypersensitivity pneumonitis, psittacosis, and Newcastle disease.

- **Poor lighting.** It is difficult for workers to see when lighting is reduced to calm the chickens. This lack of illumination contributes to slips, falls, and cuts, and makes inadequately guarded fans even more dangerous.
- **Standing for a long time.** Standing for a long time can cause pain and strains in the legs and lower back. Common types of footwear worn in this area (e.g., rubber boots) do not provide much arch support.

**Task 4: Kill Room Attendant (Backup Killer).** The kill room attendant monitors the automated poultry killing process and uses a knife to kill any birds missed by the machine. Hazards of this task may include

- **Standing for a long time.** Standing for a long time can cause pain and strains in the legs and lower back. Common types of footwear worn in this area (e.g., rubber boots) do not provide much arch support.
- **Falls, back injuries, and cuts.** Employee stands in two to three inches of blood, which creates slippery floor conditions that may result in a worker falling while holding a knife.
- **Blood on employee.** During the processing of birds in this area, blood may get in the worker's face and eyes, creating a hazard of infection and exposure to disease.
- **Restricted exit.** Employee is surrounded by equipment and product that may block sightlines where access from and to work areas is restricted. Access points may not be obvious in cases of fire or other emergencies.
- **Ergonomic hazards from use of knives.** Workers use a knife to perform this cutting task. Factors such as poorly fitting gloves, slick handles, inappropriately sized handles, or dull blades can increase the force that must be used. Repetitive or prolonged exertion of finger force when performing cutting tasks can stress the tendons and tendon sheath of the hand.

**Task 5: Picking Room Operator.** A picking room operator ensures that equipment (e.g., stunner, scalders, picks, and conveyors) functions properly. Most of the time on the job is spent walking around the equipment, performing a quality assessment of its operation. Hazards of this task may include

- **Electrical shock.** Workers in wet areas may contact electrical wires, causing electrical shock.

- **Strangulation and amputation.** Employees may need to work around moving and unguarded equipment, where an accident may result in possible amputations or strangulation.
- **Noise.** Workers in this area are exposed to high noise levels from the surrounding machinery and processing equipment, which can result in hearing loss.

**Task 6: Paw Room Grader.** The paw room grader inspects and sorts product (feet) on a conveyor. This is sometimes used as a light-duty position. Hazards of this task may include

- **Standing for a long time.** Standing for a long time reduces blood flow to the legs, forces isolated muscles to work for an extended period of time, and increases the risk of fatigue and varicose veins.
- **Prolonged bending of neck.** Employees spend a long time looking down at a conveyor belt that moves product past them. This can cause neck and shoulder pain and potentially carpal tunnel–like symptoms.

### Evisceration

Evisceration processes remove the internal organs of the poultry. Hearts, livers, gizzards, and necks may also be cleaned and packaged in evisceration. This operation includes the following tasks:

**Task 1: Rehang.** After the carcass has been removed from the kill line by cutting off the feet, it is lifted from a conveyor or shelf and rehung on shackles on the evisceration line for further processing. Hazards of this task may include

- **Reaching to the conveyor and shackles.** Employees bend and reach to lift chickens from the supply conveyor and then reach out and away, sometimes above shoulder height, to place them on a shackle conveyor. Injuries to the shoulder, back, and neck are common due to awkward postures and high repetition. Employees at the beginning of the line often work faster than those near the end of the line because there is always a full supply of birds and all shackle positions are open.
- **Standing for a long time.** Standing for a long time reduces blood flow to the legs, forces isolated muscles to work for an extended time, and increases the risk of fatigue and varicose veins.

**Task 2: Opener (Vent Opener).** The opener uses scissors to manually cut open the bird. Most companies have eliminated this position by installing an automatic vent opener machine. Employees that serve as backup to the machine monitor the birds coming out of the machine and manually open any birds that may have been missed. Hazards of this task may include

- **Ergonomic hazards from the use of scissors.** Workers often use manual scissors that can cause ergonomic stress on the arms, hands, and fingers. Repeated opening of the jaws can irritate and inflame the tendons and sheaths of the hand. This is especially a problem if employees are positioned either too high or low in relation to the bird, such that the wrist is bent while finger force is exerted. The tendon and sheath can experience contact damage as they are pulled across the bones and ligaments of the wrist. Contact between the loop handles of the scissors and the sides of the fingers can damage nerves and blood vessels.
- **Reaching to the shackles.** Workers are required to repeatedly reach to the shackles to access the bird so that various tasks can be performed. Reaching creates stress on the arms, shoulders, neck, and back because the weight of the arm and scissors must be supported.
- **Standing for a long time.**

Task 3: Neck Breaker. The neck breaker uses a knife to cut the neck of the bird. Most companies have eliminated this position by installing an automatic neck-breaking machine. Employees serve as backup to this machine. Hazards of this task may include

- **Standing for a long time.**
- **Reaching to the shackles.** Neck breakers perform various tasks by reaching repeatedly to the shackles. Reaching creates stress on the arms, shoulders, neck, and back.
- **Ergonomic hazards from use of knives.** Workers use a knife to cut the neck away from the body. The cutting motion may entail some bending of the wrist. Factors such as poorly fitting gloves, slick handles, inappropriately sized handles, or dull knives increase the force that must be used. Finger force and bending the wrist are recognized risk factors for developing many hand injuries.

Task 4: Oil Sack Cutter. The oil sack cutter cuts the oil sack from the birds. Most companies have eliminated this position by installing an automatic opening machine. Employees that serve as backup to the machine walk back and forth monitoring the procedure. Hazards of this task may include

- **Standing for a long time.**
- **Reaching to the shackles.** Workers perform various oil sack cutter tasks by reaching repeatedly to the shackles. Reaching creates stress on the arms, shoulders, neck, and back.
- **Ergonomic hazards from use of knives.** Workers use a knife to cut the oil sack. The cutting motion may entail bending the wrist. Factors such as poorly fitting gloves, slick handles, inappropriately sized handles, or dull knives increase the force that must be used. Finger force and bending of the wrist are recognized risk factors for developing many hand injuries.

**Task 5: Arranger.** The arranger, also called the presenter, removes the viscera from the body cavity and arranges them for USDA inspection. The initial removal is often accomplished by the automatic vent-opening machine. Hazards of this task may include

- **Pulling and turning viscera.** Worker repeatedly pulls the viscera from the body cavity with fingers and twists the forearm to present them for inspection. This process causes potential injury to both the wrist and elbow. The more the wrist is bent during this process, the greater the risk of injury.
- **Reaching to the shackles.** Workers perform various arranger tasks reaching repeatedly to the shackles. Reaching creates stress on the arms, shoulders, neck, and back.
- **Standing for a long time.**

**Task 6: Giblet Harvester.** The giblet harvester separates the heart, liver, and gizzard from the rest of the viscera and positions them to be cut by a saw. The heart, liver, and gizzard then fall to a wash table where an initial cleaning is performed and they are directed for further processing. Hazards of this task may include

- **Reaching to the shackles.** Workers are required to repeatedly reach to the shackles to access the bird so various tasks can be performed. Reaching creates stress on the arms, shoulders, neck, and back.
- **Standing for a long time.**

**Task 7: Gizzard Harvester.** A gizzard harvester separates gizzards from other items in the viscera and directs the gizzard to the gizzard table. Hazards of this task may include

- **Reaching across high and/or wide work surface.** Employees repeatedly reach across a conveyor or work table to obtain product for processing. Repetitive reaching stresses the shoulder and upper back and may require bending at the waist, which can stress the lower back.
- **Ergonomic hazards from use of knives.** Workers use a knife to perform some trimming and cleaning functions. Most knives have a straight, in-line design. Using this type of knife on a horizontal cutting surface forces employees to bend their wrists to perform the cut. Bending the wrist while exerting finger force is stressful to the tendons and muscles of the hand and forearm. Factors such as poorly fitting gloves, slick handles, inappropriately sized handles, or dull knives increase the force that must be used. Finger force and bending the wrist should be minimized when performing cutting tasks.
- **Ergonomic hazards from use of scissors.** Workers often use manual scissors that can cause ergonomic stress on the arms, hands, and fingers. Repeated opening of the jaws can irritate and inflame the tendons and sheaths of the hand. This is especially a problem if employees are positioned either too high or low in relation to the bird, such that the wrist is bent while finger force is exerted. The tendon and sheath can experience contact damage as they are pulled across the bones and ligaments of the wrist. Contact between the loop handles of the scissors and the sides of the fingers can damage nerves and blood vessels.
- **Standing for a long time.**

Task 8: Gizzard Table Operator. Gizzard table operators manually trim and clean gizzards. They then place gizzards in an automatic splitting machine so they are opened up and washed when they reach the gizzard peeler station. Hazards of this area may include

- **Reaching across high and/or wide work surface.** Employees repeatedly reach across a conveyor or work table to obtain product for processing. Repetitive reaching stresses the shoulder and upper back and may require bending at the waist that can stress the lower back.
- **Ergonomic hazards from use of knives.** Workers use a knife to perform some trimming and cleaning functions. Most knives have a straight, in-line design. Using this type of knife on a horizontal cutting surface forces employees to bend their wrists to perform the cut. Bending the wrist while exerting finger force is stressful to the tendons and muscles of the hand and forearm. Factors such as poorly fitting gloves, slick handles, inappropriately sized handles, or dull knives increase the force that must be used. Finger force and bending of the wrist should be minimized when performing cutting tasks.

- **Ergonomic hazards from use of scissors.** Workers often use manual scissors that can cause ergonomic stress on the arms, hands, and fingers. Repeated opening of the jaws can irritate and inflame the tendons and sheaths of the hand. This is especially a problem if employees are positioned either too high or low in relation to the bird, such that the wrist is bent while finger force is exerted. The tendon and sheath can experience contact damage as they are pulled across the bones and ligaments of the wrist. Contact between the loop handles of the scissors and the sides of the fingers can damage nerves and blood vessels.
- **Standing for a long time.**

**Task 9: Gizzard Table-Peeler Operator.** Employee presses the inside of gizzard against a rotating drum with a raspy surface that peels the inner lining from the gizzard. Employee feeds peeled gizzards to bagging area. Hazards of this task may include

- **Hands/fingers getting caught by rollers.** As workers move gizzards over peeler, fingers may get caught in rollers.
- **Reaching across high and/or wide work surface.** Employees repeatedly reach to pull product to the peeler. Repetitive reaching stresses the shoulder and upper back.
- **Standing for a long time.**
- **Prolonged and forceful finger exertion.** Employees use finger force to press gizzards against rotating drums for the majority of the task. Significant finger force must be exerted since the product is small and slippery. The wrist is usually bent during this process to place gizzards in the proper alignment. Exerting finger force for a prolonged time can stretch and fray tendons. Bending the wrist while exerting finger force can create further damage to tendons and their sheath, which can lead to injuries of the hand, wrist, and elbow.

**Task 10: Heart and Liver Cutter/Inspector.** An employee washes and visibly inspects hearts and livers before they are sent to the bagging station. Hazards of this area may include

- **Reaching across high and/or wide work surface.** Employees repeatedly reach across a conveyor or work table to obtain product for processing. Repetitive reaching stresses the shoulder and upper back and may require bending at the waist, which can stress the lower back.
- **Ergonomic hazards from use of knives.** Employees use in-line, straight knives to clean and trim. Using this type of knife on a horizontal cutting surface forces the employees to bend their wrists to perform the cut. Bending the wrist while exert-

ing finger force is stressful to the tendons and muscles of the hand and forearm. Factors such as poorly fitting gloves, slick handles, inappropriately sized handles, or dull knives increase the force that must be used. Finger force and bending of the wrist should be minimized when performing cutting tasks.

- **Ergonomic hazards from use of scissors.** Workers often use manual scissors that can cause ergonomic stress on the arms, hands, and fingers. Repeated opening of the jaws can irritate and inflame the tendons and sheaths of the hand. This is especially a problem if employees are positioned either too high or low in relation to the bird, such that the wrist is bent while finger force is exerted. The tendon and sheath can experience contact damage as they are pulled across the bones and ligaments of the wrist. Contact between the loop handles of the scissors and the sides of the fingers can damage nerves and blood vessels.
- **Standing for a long time.**

Task 11: Bagger. A bagger bulk-packs hearts, livers, or gizzards into various-sized bags before the bags are placed into boxes for shipment. Baggers may also repack giblets (heart, liver, gizzard, neck) into small paper bags for reinsertion into body cavity of whole birds. Some operations also require these employees to move the filled bag to a sealer where the bag is closed with a clip or heat seal. Hazards of this task may include

- **Repetitive pinch grips.** Employees secure and hold bag using a one- or two-finger pinch grip when removing it from the bagging fixture, transporting it to the bag sealer, and feeding it into the bag sealer. Using pinch grip places significant stress on the tendons of the fingers, which can lead to injuries of the hand, wrist, and forearm.
- **Reaching across or up to high work surfaces.** Employees repeatedly reach to bins or across table tops to obtain product for bagging and place product in bags. Repetitive reaching stresses the shoulder and upper back.
- **Standing for a long time.**

Task 12: Lung Vacuumer. A lung vacuumer uses a small suction device to remove the lungs and the kidneys from the body cavity. Hazards of this task may include

- **Awkward hand/arm postures.** Worker must bend wrist and/or pull elbow away from the body to position vacuum into body cavity to remove the lungs and kidneys.
- **Reaching to the shackles.** Workers are required to reach to the shackles to access the bird, thus causing ergonomic stress on the arms, shoulders, neck, and back.
- **Standing for a long time.**

**Task 13: Backup Eviscerator/Inspector.** The backup eviscerator is a final product inspector who feels inside the carcass and looks for any remaining pieces of viscera before removing the carcass from the shackle. Hazards of this task may include

- **Reaching to the shackles.** Workers access birds by reaching to the shackles thus causing ergonomic stress on the arms, shoulders, neck, and back.
- **Wrist deflection.** When employees are too low in relation to the bird, they must reach up to access the body cavity resulting in wrist bending. This can result in tendon and nerve damage, leading to pain and numbness in the hand, wrist, or elbow.
- **Standing for a long time.**

**Support Task: Rework Floor Person.** A rework floor person manually reworks damaged or improperly processed items. This may include trimming, washing, and salvaging parts. Often, employees receive work from tubs and then replace them onto the shackle. Hazards of this task may include

- **Bending at the waist to reach into tub.** Repeatedly bending forward and reaching out away from the body stresses the back even if there is little being lifted because the upper body must be supported. When loads are being lifted, bending over at the waist increases the distance the load is held away from the body and increases the stress placed on the back.
- **Reaching to the shackles.** Workers must reach to the shackles to place reworked product for further processing. Reaching creates stress on the arms, shoulders, neck, and back.
- **Ergonomic hazards from use of knives.** Workers use a knife to perform some trimming and cleaning functions. Most knives have a straight, in-line design. Using this type of knife on a horizontal cutting surface forces the employees to bend their wrists to perform the cut. Bending the wrist while exerting finger force is stressful to the tendons and muscles of the hand and forearm. Factors such as poorly fitting gloves, slick handles, inappropriately sized handles, or dull knives increase the force that must be used.
- **Ergonomic hazards from use of scissors.** Workers may use scissors to trim and clean product. Scissors can cause ergonomic stress on the hands and fingers, which can result in nerve and tendon damage to the hand and forearm.
- **Reaching across high and/or wide work surface.** Employees repeatedly reach across a conveyor or work table to obtain product for processing. Repetitive reaching stresses the shoulder and upper back, and may require bending at the waist, which can stress the lower back.

**Support Task: Ice Attendant.** The ice attendant manually brings ice from the ice house to the packing line, paw room, and other areas as needed. Usually, the ice is transported in tubs. Hazards of this task may include

- **Slips, trips, and falls.** Workers are standing on wet floors that may have bird skin, bird parts, and ice on them, creating a slipping hazard. Metal drain covers on the floor are also very slippery and pose a hazard. A falling worker may contact dangerous equipment.
- **Moving heavy tubs of ice.** Employees manually push tubs of ice. Pushing tubs, especially when on slick or icy floors, stresses the back, shoulder, ankle, and knee.
- **Shoveling loads of ice.** Employees support a load that can easily weigh 15 pounds from the end of a shovel handle. In a manner similar to that encountered on a child's teeter totter, leverage can increase the effect of this load by two to four times depending on the length of the shovel handle. Additionally employees may need to repeatedly bend at the waist to scoop from the bottom of the tubs and may need to lift ice above head height. The back and shoulders can be negatively affected by these motions.

### Cutting

After a chicken has been eviscerated and cleaned, it is prepared for packaging as a whole bird, or it may enter one of two processes: (1) the cutting process for preparation of a bone-in product, or (2) the cutting and deboning process for preparation of bone-out products.

In the cutting process, the wings and legs/thighs are removed from the carcass and the back is cut away from the breast. Bones are not removed. At this point parts can be packaged as a consumer product, bulk-packed for delivery to other processors, or shipped to other parts of the plant for further processing.

Removing the legs and wings from the bird is usually the beginning stage for both the packaged bone-in and bone-out product. The eviscerated birds are generally impaled on "cones" that continually pass in front of employees. Wings and the leg/thigh are cut away from the main carcass. The breast may also be cut away while on the cone. Wings and leg/thigh are sometimes removed while birds hang on a shackle conveyor similar to that used for evisceration. This operation includes the following tasks:

**Task 1: Line Loader.** Birds are often transported from the evisceration line to the cone conveyor or line in a tub. Line loaders grasp two or three birds in each hand, lift them from the tub, and place them on a conveyor or staging shelf, which is generally at waist- to shoulder-height. Other personnel usually place the birds on the cone or shackle. Hazards of the task may include

- **Bending at the waist to reach into tubs.** Repeatedly bending forward and reaching out away from the body stresses the back even if there is little being lifted because the upper body must be supported. When loads are being lifted, bending over at the waist increases the distance the load is held away from the body and increases the stress placed on the back.
- **Forceful gripping.** Employees lift multiple birds at one time, usually by the legs. Lifting two or three birds in each hand is not uncommon. Birds are cold and slick, and employees usually wear rubber gloves that are also slick and may not fit well. All these factors increase the finger force that must be exerted. Exerting significant finger force can stretch and fray the tendons of the hand and can create a contact trauma to the tendon and sheath where they come in contact with bone or tendon. These types of actions increase the risk of tendonitis and carpal tunnel syndrome.

Task 2: **Tail Cutter.** The tail is cut from the bird before the bird is placed on the cone. A standard scissors is generally used to perform the operation. Hazards of this task may include

- **Ergonomic hazards from use of scissors.** Using traditional scissors forces the fingers to repeatedly open and close the blade, which can stress tendons, increasing the risk of tenosynovitis and carpal tunnel syndrome. Contact trauma to sides of fingers can damage nerves, which can cause numbness and tingling in the tips of the fingers and thumb.
- **Standing for a long time.** Standing for a long time reduces blood flow to the legs, forces isolated muscles to work for an extended time, and increases risk of fatigue and varicose veins.

Task 3: **Saw Operator.** Employees may use a saw with a manual feed to cut leg/thigh or wings away from the main carcass, or may load a machine that automatically performs cuts. Manual feed saws can be used to remove legs from the back, divide the legs, cut wings away from the breast, and split the breast in two. After being loaded, automated machines perform the same cuts as described above. Hazards of this task may include

- **Reaching to access product, saws, or machine load areas.** Employee reaches repeatedly to conveyor or shelf to obtain birds for processing. Reaches are also nec-

essary to place birds into the automatic saw feed mechanism and perform manual cuts. Repetitive reaching stresses the shoulder and upper back.
- **Cuts and lacerations.** The nature of this task involves employees working with unguarded saws. Cuts, lacerations, and amputations are possible.
- **Standing for a long time.**

**Task 4: Rehang.** Rehang is generally not necessary, since most cutting is performed on a cone line. If the cutting is to be performed from a shackle conveyor, the bird must be rehung. Some automated cutters, such as a "multi-cut" machine, must be loaded, and this is technically a rehang type of activity. The bird must be lifted from the table or conveyor and the legs placed into a shackle or other device moving in front of the employee. This is a highly repetitive reaching task. Hazards of this task may include

- **Reaching up, forward, or to the side to access the shackle.** Employees may bend to lift chickens from the supply conveyor and then reach out and away, sometimes above shoulder height, to place them on multi-cut machines or shackle conveyors. Injuries to the shoulder, back, and neck are common due to awkward postures and high repetition. Employees at the beginning of the line often work faster than those near the end of the line because there is always a full supply of birds and all positions are open.
- **Standing for a long time.**

**Task 5: Cone Line Feeder.** Most plants use a cone line as the main staging area for removing appendages and meat from the body of the bird. The feeder places the eviscerated carcass onto the cone, which is integrated into a conveyor line. This line moves the bird past employees who remove parts from the carcass.

In some plants parts are removed from birds hanging from a shackle conveyor, or the process may be automated using multi-cut machinery. In these cases the cone line feeder is replaced by a rehang worker. Hazards of this task may include

- **Reaching.** Employees repeatedly reach to a conveyor or shelf to obtain birds for processing and reach to place birds on the cone. Repetitive reaching stresses the shoulder and upper back.
- **Standing for a long time.**

**Task 6: Wing Cutter.** Wing cutters use knives to cut the wings from the bird. This may be a multi-step process where several workers along the line each perform one of the necessary cuts, or all cuts can be done by a single operator. Hazards of this task may include

- **Ergonomic hazards from use of knives.** Workers use a knife to cut the wings away from the rest of the carcass. The cutting motion may entail some bending of the wrist. Factors such as poorly fitting gloves, slick handles, inappropriately sized handles, or dull knives increase the force that must be used. Finger force and bending of the wrist are recognized risk factors for the development of many hand injuries.
- **Cuts and lacerations.** Employees are performing highly repetitive tasks using knives close to other employees. Cuts and lacerations are possible to the employee and those standing nearby because employees are exposed to sharp knife blades. Any cut not treated at once will normally become infected as a result of working with poultry.
- **Reaching.** Employees repeatedly reach to the bird on the cone to perform cutting tasks and may need to reach to a bin or a tub to deposit removed item. Repetitive reaching stresses the shoulder and upper back.
- **Standing for a long time.**

**Task 7: Leg/Thigh Cutter.** Cutters use knives to cut the legs/thigh unit from the bird. This may be a multi-step process where several workers along the line each perform one of the necessary cuts, or all cuts can be done by a single operator. Hazards of this task may include

- **Ergonomic hazards from use of knives.** Workers use a knife to cut the wings away from the rest of the carcass. The cutting motion may entail some bending of the wrist. Factors such as poorly fitting gloves, slick handles, inappropriately sized handles, or dull knives increase the force that must be used. Finger force and bending of the wrist are recognized risk factors for the development of many hand injuries.
- **Cuts and lacerations.** Employees are performing highly repetitive tasks using knives close to other employees. Cuts and lacerations are possible to the employee and those standing nearby because employees are exposed to sharp knife blades. Any cut not treated at once will normally become infected as a result of working with poultry.
- **Reaching.** Employees repeatedly reach to the bird on the cone to perform cutting tasks and may need to reach to a bin or a tub to deposit removed item. Repetitive reaching stresses the shoulder and upper back.

**Task 8: Back/Breast Separator.** Employees may use a saw with a manual feed to separate the breast section from the back. This manual feed technique can be used to remove legs from the back, divide the legs, cut wings away from the breast, and split the breast in two. After being loaded, automated multi-cut machines perform the same cuts as described above. Hazards of this task may include

- **Reaching to access product, saws, or machine load areas.** Employee repeatedly reaches to conveyor or shelf to obtain birds for processing. Repetitive reaching stresses the shoulder and upper back.
- **Cuts and lacerations.** The nature of this task involves employee working with an unguarded saw. Cuts, lacerations, and amputations are possible.
- **Standing for a long time.**

**Task 9: Trimmer/Clean-up.** Employee obtains separated pieces of poultry from conveyor and uses scissors to trim excess skin, fat, and pieces of bone. Hazards of this task may include

- **Ergonomic hazards from use of scissors.** Use of traditional scissors forces the fingers to repeatedly open and close the blade, which can stress tendons, increasing the risk of tenosynovitis and carpal tunnel syndrome. Contact trauma to sides of fingers can damage nerves, which can cause numbness and tingling in the tips of the fingers and thumb.
- **Standing for a long time.**
- **Reaching hazard.** Employees repeatedly reach to conveyor or shelf to obtain parts to be trimmed and reach to place finished parts in tubs or baskets. Repetitive reaching stresses the shoulder and upper back.

**Support Task: Knife Person.** A knife person collects dull knives from employees along the processing lines and replaces them with sharp ones. This employee may also sharpen knives that have been collected. Hazards of the task may include

- **Slips, trips, and falls.** Workers walk all over the facility on wet floors that may have bird skin, bird parts, and ice on them, creating a slipping hazard. Metal drain covers on the floor are also very slippery and pose a hazard. A falling worker may contact dangerous equipment, or cut himself or herself on a knife blade.
- **Hazards from use of grinders.** Employees may suffer cuts, lacerations, skin abrasions, contusions, or eye damage during use of grinders to sharpen knives.

Grinding wheels may break up or explode. Bits and pieces of knife blades may be thrown off during sharpening.

### Deboning

Within-plant processing of cut-up parts generally involves the creation of a bone-out product. The deboning process involves cutting meat away from the bone using traditional knives or Whizzard knives and trimming and cleaning with traditional bladed knives or scissors. The deboned parts are generally packaged as a fresh or flash-frozen consumer product.

DID YOU KNOW?

In some facilities, the birds are aged in a cooler before the meat is separated from the bone. (OSHA 2007c)

All parts of the chicken that have sufficient meat are candidates for a bone-out product. The deboning process generally follows the cutting operation when deboning legs and thighs. The breast meat, however, may be removed while the carcass is still on the cone at the end of the cutting process.

While most deboning cuts are performed with a knife, some of the processes, such as removal of the meat from the bone of the leg, can be more easily and safely accomplished by using other cutting tools such as a Whizzard knife. Trimming tasks in this process are generally performed with scissors.

After the meat has been removed from the bone it may be quick-frozen and bagged with little additional trimming or processing, or it may be moved to a separate specialty trim line where it is trimmed and cut according to customer requirements. This operation includes the following tasks:

**Task 1: Skin Puller.** Employee uses pliers or similar tool to pull skin from breasts, thighs, and legs. Hazards of this task may include

- **Standing for a long time.** Standing for a long time reduces blood flow to the legs, forces isolated muscles to work for an extended time, and increases risk of fatigue and varicose veins.
- **Forceful hand exertions.** Employees exert high finger force with both hands to open and close the tool and to hold the product while the skin is being pulled.

Using gloves and handling cold product increases the amount of finger force that must be exerted. Repeatedly exerting high finger force can stretch and fray the tendon if there are not sufficient periods of rest. Repeatedly stretching the tendon can lead to tendinitis or tenosynovitis. Using standard in-line tools can cause employees to bend the wrist, which, in combination with high finger force, can cause contact trauma between the tendon and the bones and ligaments of the wrist. Contact between these entities can cause irritation and inflammation, leading potentially to tendinitis, tenosynovitis, and carpal tunnel syndrome.

**Task 2: Line Loader.** Parts are transported from the cutting stations to the deboning stations in hand-carried tubs. Tubs may be lifted from cart shelves or lifted from the floor and carried to deboning stations. Tubs are emptied onto a conveyor or staging shelf which is generally at waist- to shoulder-height. This task may also be repeated at the trimming lines where product that has been deboned may again be placed in a tub and transported to the trimmers. Hazards of this task may include

- **Bending at the waist to lift tubs of product.** Repeatedly bending forward and reaching out away from the body stresses the back even if there is little weight being lifted because the upper body must be supported. When loads are being lifted, bending over at the waist increases the distance the load is held away from the body and increases the stress placed on the back. Bending and lifting heavy loads such as those encountered at these stations greatly increases the risk of injury to the low back.

**Task 3: Deboner.** Employees remove bones from various poultry parts, including breasts, thighs, and legs. The task may be performed with a standard knife on a flat cutting surface or a tilted cutting surface. Legs may be deboned on a specialized conveyor line using a Whizzard knife. Breast meat may be removed directly from the carcass while it is still on the cone line. Hazards of this task may include

- **Ergonomic hazards from use of knives.** Workers use a knife to cut the meat away from the bone. Most knives have a straight, in-line design. Using this type of knife on a horizontal cutting surface forces employees to bend their wrists to perform the cut. Bending the wrist while exerting finger force is stressful to the tendons and muscles of the hand and forearm. Factors such as poorly fitting gloves, slick handles, inappropriately sized handles, frozen meat or dull knives increase the force that must be used.

- **Cuts and lacerations.** Employees are performing highly repetitive tasks using knives close to other employees. Cuts and lacerations are possible to the employee and those standing nearby because employees are exposed to sharp knife blades. Any cut not treated at once will normally become infected as a result of working with poultry.
- **Reaching.** Employees repeatedly reach to a conveyor or shelf to obtain parts for deboning and reach to place finished product in tubs or receptacles. Repetitive reaching stresses the shoulder and upper back.
- **Standing for a long time.**

   **Task 4: Tender Puller.** Tender pullers use their fingers to pull tenders away from the breast bone after the main section of breast has been removed. This task may be performed while the carcass is on the cone line or from a cut full breast placed on a flat work source. Hazards of this task may include

- **Reaching.** Employees repeatedly reach to conveyor or shelf to obtains breasts, reach to cones to pull tenders, or reach to place finished product in tubs or receptacles. Repetitive reaching stresses the shoulder and upper back.
- **Standing for a long time.**

   **Task 5: Trimmer.** Trimming is usually the last processing step before packaging or quick freezing. Trim lines often produce specialty products according to customer specification. Trimmers remove pieces of bone, fat, tendons, gristle, or blemishes in the meat as well as perform specialty cutting to produce tenders and nuggets. Many of the items that are easily grasped, such as bone and fat, can be pulled away from the meat using only the fingers. Although a knife can be used, for the majority of these operations the tool of choice is usually scissors. Hazards of this task may include

- **Ergonomic hazards from use of scissors.** Use of traditional scissors forces the fingers to repeatedly open and close the jaws, which can stress tendons, increasing the risk of tenosynovitis and carpal tunnel syndrome. Contact trauma to the sides of fingers can damage nerves, which can cause numbness and tingling in the tips of the fingers and thumb.
- **Standing for a long time.**
- **Reaching.** Employees repeatedly reach to a conveyor or shelf to obtain parts for trimming and reach to place finished product in tubs or receptacles. Repetitive reaching stresses the shoulder and upper back.

   **Task 6: Quality Control Inspector.** Employee pulls selected processed parts from conveyor line and visually inspects for compliance with quality standards. Hazards of the task may include

- **Reaching.** Employees repeatedly reach to a conveyor or shelf to obtain parts for inspection. Reaches are also necessary to place inspected parts in tubs or back on the line. Repetitive reaching stresses the shoulder and upper back.
- **Standing for a long time**
- **Support Task: Knife Person.** See under "Cutting," above.

### Packout

Packaging is necessary to get the processed product from the plant to the consumer. It is generally a two-part procedure. First, the bird or bird parts are placed in a bag or package; and second, the package is placed in a shipping box. Poultry can be packaged in a wide variety of formats, which range from minimal processing to maximum processing:

- **Whole-bird bulk packaging.** The whole bird can be bulk boxed and sent to large users such as broiler restaurants or secondary processors.
- **Whole-bird individual packaging.** The whole bird is individually bagged and boxed for supermarket sale.
- **Bone-in product.** Parts are packaged and sold as consumer product or as bulk sale for large commercial users.
- **Bone-out product.** Parts are packaged and sold as consumer product or as bulk sale for large commercial users.

No matter how a bird is packaged, it is almost always placed in a large cardboard box for shipping. The steps in building, filling, weighing, and stacking these boxes are almost always the same. This operation includes the following tasks:

**Task 1: Box Maker.** The box maker is responsible for building boxes out of flat box stock and providing them to the stations where they will be filled with product. This task may be automated. Hazards of this task may include

- **Congested work area.** Problems can occur at this task because it is often an unplanned workstation, originally intended to be temporary. Placing such a workstation in a process line without adequate planning and space allocation can result in a highly congested work area, creating the potential for injury from working in awkward postures and coming in contact with other moving machinery.
- **Hot glue burns, splashes.** Problems can occur at this task station from exposure to hot glue as boxes are made. There is potential for employee burns and splashing hot glue when adding glue sticks to the pot, or when cleaning the machine.
- **Poor access to emergency exits.** Box-making operations are often placed in remote and hard-to-get-to locations, which makes access to emergency exits difficult. Quick

and easy access to emergency exits from any room or workstation is very important to the safety of the employee. Without it, employees could be trapped or killed if unable to reach emergency exits during an emergency.

- **Bending and reaching.** Employees are required to repeat the same motion over and over (i.e., reaching and/or bending to obtain box stock and placing finished boxes on conveyor), which can result in work-related musculoskeletal disorders.

**Task 2: Box Packer.** Packing boxes involves taking packed trays, bags, or whole birds from the conveyor and depositing them in a box. Employees generally work at a boxing station located next to a conveyor that supplies product to be boxed. Hazards of this task may include

- **Repetitive reaching and lifting.** Employees repeatedly reach to conveyor to obtain product for processing and may reach to place product in the box. Repetitive reaching stresses the shoulder and upper back.
- **Standing for a long time.** Standing for a long time reduces blood flow to the legs, forces isolated muscles to work for an extended time, and increases risk of fatigue and varicose veins.

**Task 3: Scale Operator.** A scale operator pulls filled boxes from a conveyor and places them onto scales. The operator then adds or removes product until the desired weight is achieved. This employee usually affixes a label showing the box weight before placing the weighed box back on the conveyor. Hazards of this task may include

- **Lifting heavy loads.** Employees may experience back strain from reaching, twisting, and bending if the task requires them to lift boxes, which may weigh 40–80 pounds, from the conveyor to the scales.
- **Standing for a long time.**

**Task 4: Box Sealer.** After weighing takes place, the box sealer adds ice as required and places a lid on the box. Sealer pushes box down conveyor to stack-off employee. Hazards of this task may include

- **Standing for a long time.**

**Task 5: Stack Off.** The stack-off employee is responsible for removing boxes, which may weigh 40–80 pounds, from the conveyor and stacking them on pallets. Palletized loads are then stored or loaded onto trucks. Hazards of this task may include

- **Bending and twisting while lifting heavy loads.** Stacking full boxes forces employees to bend the torso forward to place the box on the pallet, or to lift the box up to or above head height to place it on the top of the stack. This can result in back injuries, since boxes can weigh up to 80 pounds.
- **Slips, trips, and falls.** Employees must often move pallets of product with a hand jack while stepping on wet, slippery surfaces, resulting in possible slips and falls.

### Warehousing

Once the chicken is packed in its shipping container, it is moved from the processing floor. Options are moving to a truck for immediate shipment or placement in a warehouse for storage. Lifting and moving heavy loads using awkward body postures is common practice. This operation includes the following tasks:

**Task 1: Forklift/Pallet Jack Operators.** Employees move heavily loaded pallets of poultry around the worksite and place them onto trucks for shipping using forklifts or pallet jacks. Hazards of this task may include

- **Collision.** Forklifts are driven in heavily congested areas where many other operations are being performed. Visibility and sightlines can be limited, increasing the chance of collisions between machinery and employees.
- **Defective forklift.** Over long periods of time the solid rubber wheel on the front of the forklift has developed a flat spot, resulting in unstable loads and poor handling.
- **Struck by falling loads.** There is a potential for pallets to collapse or tip on driver; uneven floor areas increase the danger.
- **Fumes from battery charging.** Forklift batteries need to be removed by hoist and routinely charged. Exposure to fumes or vapors from battery charging process can be dangerous.
- **Hazards from working with pallet jacks.** While using hand or electric jacks, employees may have the problems of backing into walls, running over own feet, or running into other employees as they move through freezer doors. It is difficult to see who or what is on the other side of the doors, because the windows may be high and covered with frost.

**Task 2: Freezer/Cooler Worker:** Employees manually load and unload boxes of frozen product on warehouse racking shelves. Occasional palletizing of loads may be necessary. Hazards of this task may include

- **Bending and twisting while lifting heavy loads.** Stacking boxes over head height forces employees to bend the torso forward to place the box on the pallet, or to lift the box up to or above head height to place it on the top of the stack. Even when boxes are moved by conveyors directly to refrigerated trucks, employees remove the boxes from the conveyor and then stack them in the truck above head height or at floor level. This can result in back and shoulder injuries.

- **Exposure to cold environments.** Employees working in cold environments must wear additional clothing, which can restrict movement and increase the force they must exert when performing lifting operations. Additionally, employees burn more energy in these environments to keep warm so fatigue may occur more rapidly, which increases the risk of injury. Cold areas may also have ice forming on work surfaces where slipping can cause injury or strain, especially if a load is being carried when the slip occurs (OSHA 2007c).

## SAFE WORK PRACTICES IN THE POULTRY INDUSTRY

As shown in the preceding sections, and from poultry industry on-the-job injury, illness and fatality statistics, a number of operations, conditions, or practices in the industry can be considered to pose hazards. As mentioned, the list includes tasks that could result in cuts or lacerations, repetitive motion disorders, slips and falls, exposure to cold and wet environments, exposure to dust, dermatitis, exposure to chemicals, and noise exposure. The remainder of this chapter reexamines many of these previously mentioned potential problems in the poultry industry and suggests preventive measures and possible solutions. These solutions from the findings of the North Carolina Department of Labor (NCDOL 2005). Based on personal experience, we have found that finding and defining a safety and health hazard is normally the easy part of any investigation; coming up with correct mitigation actions to prevent or lessen the possibility of injury to workers on the job is not as simple.

### Hazard: Cuts and Lacerations

As mentioned, saws, knives, and scissors are the tools used in cutting and deboning chicken. Saws are used to cut chicken into quarters and pieces. In the deboning process, the chicken carcass is placed on a cone on a conveyor line. At each work station a different cut is made, to remove the legs, wings, skin, breast meat, and thighs. Each operator on the line makes one or more specific cuts with a very sharp knife to remove specific portions of meat. Scissors are used to trim bone, gristle, and fat from the meat. Tools that are sharp enough to cut meat will easily cut fingers.

### *Preventive Measures*

Personal protective equipment (PPE) such as metal mesh gloves and arm guards will help to reduce the number of cuts or lacerations workers may experience in the deboning process. Employees can be trained in methods and techniques that produce clean cuts and prevent injuries. Knives and scissors must be kept very sharp so that the appropriate cut can be made. Sharp tools also help reduce the force required to make the cut. Accidental injuries can be reduced by keeping knives and scissors in scabbards when not in use. Sufficient space between operators will help prevent employees from accidentally cutting each other.

Saws used to cut birds into quarters should have appropriate guards on the blade to protect the operator from injury. Adjustments to the saw must be accomplished when the power is off and the machine is stopped. Particular saws should have a lower guard that retracts when the saw is in use, then automatically returns to the guard position. Appropriate grounding and insulation of the saws (and all electrical equipment) are necessary to prevent electrical hazards (NCDOL 2005).

## GROUND FAULT

An unintentional, electrically conducting connection between an ungrounded conductor of an electrical circuit and the normally non-current-carrying conductors, metallic enclosures, metallic raceways, metallic equipment, or earth.

## GROUND-FAULT CIRCUIT INTERRUPTER

A device intended for the protection of personnel that functions to de-energize a circuit or portion thereof within an established period of time when a current to ground exceeds the values established for a Class A device. (A Class A ground-fault circuit interrupter trips when the current to ground has a value in the range of 4 mA to 6 mA.) (NFPA 2004)

Minor cuts and lacerations should immediately be thoroughly washed with soap and water and treated with an antiseptic and dressing. Deep cuts or lacerations with loss of motion to the affected area should be referred to a doctor for treatment. Prompt treatment will help reduce infection and promote early healing. All injuries should be reported to supervision.

**Hazard: Dermatitis**

Skin disorders, or dermatitis, may be the most frequently occurring occupational illness. Conversely, the skin, as the largest organ of the human body, is one of our most valuable weapons in preventing illnesses. For example, the skin allows us to cope with extremes in our environment, including temperature, moisture, wind, and weather. Yet responses prompted by the skin to the season of the year sometimes contribute to skin diseases. In warmer weather workers tend to wear less clothing so that there is greater likelihood of skin contact with irritants. In cold weather there is more potential for chapping from exposure to cold and wind. Overheated homes and workplaces can cause skin to become dry and more easily damaged.

Skin disorders can affect any worker, regardless of the type of job or the industry. Most occupational skin disorders begin on exposed skin such as hands or arms. Skin disorders include callouses and blisters caused by pressure and/or friction. Other skin disorders are burns and frostbite, which are caused by extremes of heat or cold. Biological agents such as plants or animals or bacteria can cause dermatitis.

Skin disorders range from red, chapped hands to lesions and eruptions. People with preexisting skin disease are more at risk of developing occupational dermatoses. Skin disease unrelated to one's present job can be aggravated by exposures at work. Individuals with skin allergies may react to very small amounts of a substance to which they are allergic.

Some employees who work in poultry processing develop skin rashes and dermatitis. This may be caused by contact with the water used to clean and rinse the chicken during preparation of the birds. It is important to wash hands and arms frequently with soap and water and to dry them thoroughly. Protective clothing such as gloves may be required for some operations. Individuals with preexisting conditions should not work where the disorder could be aggravated. Because people who work in the poultry preparation department frequently expose their hands to temperature changes, chapped skin may be experienced.

*Preventive Measures*

Personal cleanliness is the most important measure for preventing skin irritations or rashes. Thorough washing and drying of hands, arms, and other involved skin areas are essential.

The use of hand creams or water-repellent protective barrier creams may help in the prevention of occupational dermatitis. Several times during the workshift, the creams should be removed by washing with soap and water. The skin should be dried, and the creams should be reapplied.

Particular jobs may require the use of rubber gloves to keep the skin out of contact with water. Protective equipment should fit correctly to prevent additional irritation from too tight a fit or from rubbing when the fit is too loose. Protective equipment should be inspected frequently, kept in good repair, and replaced when necessary.

### Hazard: Cold/Wet Environment

As birds move along in processing, they are washed and cooled with cold water. Temperatures in the poultry processing facilities are kept cool to prevent meat spoilage and to conform to USDA requirements that meat of the chicken be kept at 40°F. This produces a work environment that is wet and cold.

Workers with poor circulation to the extremities (hands and/or feet) may experience increased discomfort in a cold and wet environment because of the additional constriction of blood vessels caused by the cold. Long-term exposure to a cool, damp environment also produces more discomfort for individuals with musculoskeletal disorders such as arthritis. Other chronic diseases that affect the nerves and blood vessels on the hands or feet can be aggravated by cold and wet work areas.

As previously noted, the cool and wet environment required for chicken processing contributes to chapping of the skin and may aggravate skin disorders.

*Preventive Measures*

Workers should wear warm clothing. Where possible, protective clothing such as rubber aprons should be worn to keep clothing dry. Rubber boots with heavy socks will keep the feet warm and dry. In some areas where there is frequent handling of very cold meat, cotton gloves can be worn under rubber gloves to keep the hands warm.

### Hazard: Slips and Falls

Floors and work areas in poultry processing will be wet because of the wet process and frequent cleaning required for sanitary reasons. Similarly, grease or fat from the birds will make floors and work areas slippery. Standing and walking in work areas with slippery floors increases the potential for slips and falls.

*Preventive Measures*

Good drainage is essential for work areas where there is wet processing. Boots and nonskid soles and floor mats with nonskid surfaces can be used to reduce the potential for slips and falls. Routinely scheduled cleaning during the workshift helps maintain a sanitary work environment and reduces the buildup of grease and fat. Aisles and passageways where mechanical handling equipment is used should be clearly marked and be of sufficient size for safe clearance.

### Hazard: Respiratory Irritants

Poultry industry employees can be exposed to a variety of respiratory irritants. Chicken handlers or chicken growers experience the greatest exposure to airborne contaminants because of dust from feed grains, gases from decomposing manure, dander, and feathers. The level of air contaminants increases with older birds. Dust and gas levels are higher in colder months because buildings are open during warmer weather.

In the poultry processing operations, exposure to respiratory irritants is heavier in the receiving area because of the activity of excited, nervous birds. Exposure is heaviest when birds are removed from their cages and live hung on conveyor lines. Individuals who are exposed to these respiratory irritants may have symptoms such as cough, shortness of breath, wheezing, stuffy nose, or eye irritation.

There are other respiratory illnesses that are uncommon diseases but can occur from infectious pathogens. Some of these are believed to spread to humans from infected vapors from evisceration in poultry processing. Symptoms include fever, chills, aching muscles, headache, and inflammation of the lungs.

Once the birds are killed and plucked, the potential for exposure decreases rapidly. After the birds are through the preparation areas and move to processing, there is no longer a problem with airborne respiratory irritants.

### Preventive Measures

Insofar as possible, the receiving area should be kept clean and free from trash. Workers who unload trucks and crates and live hang the birds should wear personal protective equipment such as nuisance dust respirators. Protective clothing and

### DID YOU KNOW?

Airborne contaminants usually measure well below OSHA limits in poultry processing operations. Nonetheless, employees who are hypersensitive to respiratory irritants in the poultry processing environment will have difficulty working there. (NCDOL 2005)

boots will help minimize exposure in this area. Eye protection is useful to prevent particles of dust and dirt from getting into the eyes. As the receiving area is primarily mechanized, good housekeeping practices and protective equipment are usually adequate to prevent problems associated with dust exposure. Workers with sensitivity to airborne contaminants should not work in this area (NCDOL 2005).

### Hazard: Chemical Exposures

There are some areas in a poultry processing plant where workers may be exposed to harmful levels of chemicals or air contaminants. Where there are high levels of chemicals in the work area, engineering controls must be used to prevent or limit employee exposure to them. Protective equipment must be used as appropriate.

Carbon dioxide in the form of dry ice is used in many poultry processing plants to keep meat cold while in holding areas and to quick-freeze meat for shipping. Carbon dioxide is an odorless gas. Inhalation of high levels of carbon dioxide may cause an increase in the breathing rate, which can progress to shortness of breath, dizziness, or vomiting.

Ammonia or Freon may be present in a poultry processing plant as chemicals used for refrigeration. Ammonia may cause irritation of the respiratory tract and the eyes. Freon is hazardous only at extremely high exposure levels. Periodic monitoring of the work areas will help detect these chemicals before they can cause a serious health effect.

Chlorine is sometimes added to the water used for washing chickens. In a diluted form chlorine is a disinfectant and usually does not present a hazard. In a concentrated form, chlorine is a respiratory irritant that can cause breathing difficulties. Even at low levels, prolonged exposure to chlorine may cause skin irritation for some workers.

Solvents or compounds used in cleaning and degreasing operations are potential health hazards. These materials are used in the maintenance of equipment and in housekeeping. When such materials are used improperly, there is the potential for inhalation of vapors that may cause a lack of coordination or drowsiness. Workers showing these symptoms should be immediately removed to fresher air.

Skin contact with cleaning compounds and solvents may cause dermatitis, ranging from simple irritation to skin damage. Products designed to remove fat and grease from equipment will also remove the natural oil barrier from the worker's skin, leaving the skin unprotected. Workers who use such compounds should be required to use appropriate protective equipment, such as gloves or barrier creams. Cleaning compounds and solvents may pose the hazard of being splashed into employees' eyes. Where strong concentrations or caustic solution agents are used, protective eyewear should be used and eye wash stations should be readily available.

Vats, tanks, or other enclosed spaces that may contain organic matter (skin, feathers, fat, offal, and so forth) should be tested for the presence of hydrogen sulfide or methane produced by the decomposing organic materials. When entering a confined space, specific work practices must be instituted and respirators must be worn. Ventilation controls should be in place.

### Preventive Measures

All workers who have the potential for exposure to chemicals should be medically evaluated prior to working in areas where they may be exposed to chemicals. Work areas that may expose workers to chemicals should be frequently monitored to detect exposure levels. Good ventilation and airflow is usually sufficient to protect workers from harmful exposure. Heavier concentrations may require local exhaust ventilation. When monitoring indicates the possibility of overexposure, workers should use appropriate respirators. Such respirators must be fitted to each individual, and employees must be trained to use the respirators.

Workers who handle chemicals such as dry ice should wear gloves to protect their hands and fingers from frostbite. Gloves and barrier creams can protect workers by preventing skin disorders from frequent contact with chemicals. Any time there is the potential for exposure to strong concentrations of chemicals, planned work practices should be followed to prevent overexposure. Where strong concentrations and caustic agents are used, protective eyewear should be worn and eye wash stations should be readily available.

Any workplace that uses hazardous chemicals is required to have a written hazard communication program in place. Among other things, the hazard communication program requires a label and material safety data sheet (MSDS) for each hazardous chemical. The label and MSDS must provide numerous specified items of information to employees. Additionally, employee training is required. The hazard communication program is required by the Hazard Communication Standard, 1910.1200, which defines a hazardous chemical as "any element, chemical compound, or mixture that is a physical hazard or a health hazard."

### Hazard: Noise

Excessive noise can cause permanent hearing damage. OSHA standards require employers to maintain a hearing conservation program when employee exposure to noise is at or above an eight-hour TWA (time weighted average) of 85 dBA (decibels). The hearing conservation program requires noise or sound level monitoring. If the exposure is above 90 dBA, the program requirements include administrative or engineering controls to try to reduce exposure to or below 90 dBA.

Annual audiograms are required for all employees exposed at or above 85 dBA. For exposure between 85 and 90 dBA, use of hearing protection is required under certain conditions. Hearing protection is required if administrative and engineering controls do not reduce the exposure to 90 dBA or less.

*Preventive Measures*

There are several ways to protect against potentially harmful noise sources. Engineering controls, such as muffling noise by enclosing noisy equipment or moving parts, are preferred. Administrative controls, such as the rotation of workers, may in some instances be used to limit the amount of time each individual is exposed to high levels of noise. Hearing protectors may be necessary to attenuate noise levels. Whenever hearing protectors are used, each employee must be provided a choice of protector, must be individually fitted, and must be trained to use the protector(s) (NCDOL 2005).

## REFERENCES AND RECOMMENDED READING

Best Food Nation. 2007. America's Poultry Industry. bestfoodnation.com/poultry.asp (accessed August 4, 2007).

Occupational Safety and Health Administration (OSHA). 1994. Hazard Information Bulletin on Contracting Occupationally Related Psittacosis. www.osha.gov/dts/hib/hib_data/hib19940808.html (accessed January 2008).

———. 2007a. Fire Safety. www.osha.gov/SLTC/etools/poultry/general_hazards/firesafety.html (accessed August 5, 2007).

———. 2007b. Poultry Processing. www.osha.gov/SLTC/poultryprocessing/index.html (accessed August 4, 2007).

———. 2007c. Poultry Processing Industry eTool. www.osha.gov/SLTC/etools/poultry/general_hazards.html (accessed August 5, 2007).

National Fire Protection Association (NFPA). 2004. *NFPA 70E Standard for Electrical Safety in the Workplace.* Quincy, MA: National Fire Protection Association.

North Carolina Department of Labor (NCDOL). 2005. *A Guide to Safe Work Practices in the Poultry Processing Industry.* Raleigh, NC: North Carolina Department of Labor. www.nclabor.com/osha/etta/indguide/ig34.pdf (accessed August 5, 2007).

# 5

# Preserved Fruits
# and Vegetables Industry

## NATURE OF THE INDUSTRY

The preserved fruits and vegetables industry includes not only canned and frozen fruits and vegetables but also canned and frozen specialties, dried and dehydrated fruits and vegetables, pickles, and salad dressings. It is important to note that the products of this industry are distinguished by their processing rather than by their container, and the term "canned products" often refers to items that are not actually canned.

Food processing occupies a powerful position within the food and fiber system. The industry has been likened to the center of an hourglass: raw agricultural commodities from more than two million farms and ranches flow through roughly 20,000 processors, which in turn sell their array of processed products to more than half a million food wholesalers and retailers. Over a hundred million domestic households consume the meat and dairy products, canned and frozen fruits and vegetables, milled grains, bakery products, beverages, and seafood.

The importance of food processing lies in its various economic functions. Foremost, processors convert food materials into finished, consumer-ready products using labor, machinery, energy, and management. They employ handling, manufacturing, and packaging techniques to add economic value to raw commodities harvested from the farm or the sea. Virtually all agricultural products are processed to some degree before reaching consumers. The value added varies by commodity: steers become meat, potatoes are turned into french fries, wheat is made into flour, apples become juice or sauce, and fresh salmon emerges as canned salmon. The farm value of fruit and vegetable products at the retail level—frozen peas, for instance—is about 20 percent. Thus, 80 percent of the retail value is "added" to the raw product during processing and distribution.

Processors serve as middlemen within the food system. Consumer demand and agricultural supply information come together at the food processing center. For instance, a tight supply of frozen corn at the retail level is eventually translated into higher processor prices, a greater willingness to pay for key inputs, and a price signal to farmers to expand production or sell off their stored crop. In contrast, an unexpectedly short crop induces processors to raise their prices to retailers and distributors, which subsequently prompts a decrease in consumer demand (OSHA 2007).

The EPA supplies more detailed information about fruits and vegetables:

> The canning of fruits and vegetables is a growing, competitive industry, especially the international export portion. The industry is made up of establishments primarily engaged in canning fruits, vegetables, fruit and vegetable juices; processing ketchup and other tomato sauces; and producing natural and imitation preserves, jams, and jellies (EPA 1995a).

Along with industry growth, technology and automation have had a large impact on the fruit and vegetable processing business. For instance, much of the manual, labor-intensive work of the past has been replaced by machines and computer-operated product processing programs that control machine operation. This does not mean that all humans in the industry have been replaced by machines and computers. Instead, it means that the total number of humans employed today has been reduced to perform tasks that in the past required a much larger workforce.

Again, it is important to point out that the food manufacturing industry is marked by a high rate of on-the-job injury and/or illness. This is the case even though, as mentioned, significant technological upgrades and computerized operation have been implemented. Even with the reduced number of workers per plant, the fruit and vegetable processing industry continues to contribute to this high on-the-job injury/illness rate. Many of the potential hazards and their sources in the fruit and vegetable processing industry are the same as many of the other sectors making up the food processing industry as a whole. For example, typical hazards include being struck by falling objects; being caught in point of operation in conveyors; slips, trips, and falls; being struck by flying objects from box staple machines; contact with toxic or noxious substances released from lift trucks and other sources; and noise generated from conveyors and other machinery.

To be able to recognize and understand the extent of potential hazard exposure in the fruit and vegetable processing industry, it is important to be familiar with the actual fruit and vegetable preservation (production) process. This is the approach the U.S. EPA (1995a) takes in the canned fruits and vegetable process description that we include below. Keep in mind that the EPA primarily points out potential emissions hazards affecting the environment and workers involved in the processing procedures. While certain process emissions can be hazardous both to the environment and to workers' safety and well-being, it is our purpose in this text to provide an overview of

**DID YOU KNOW?**

According to OSHA, more than 46 percent of injuries and illnesses reported from 1995 to 2000 in the fruit and vegetable preservation industry consisted of sprains, strains, and bruises. (OSHA 2007)

several different exposures; that is, to overview potential safety and health hazards that could be present during the processing of fruits and vegetables.

## FRUIT AND VEGETABLE PRESERVATION PROCESS

The primary objective of food processing is the preservation of perishable foods in a stable form that can be stored and shipped to distant markets during all months of the year. Processing also can change foods into new or more usable forms and make foods more convenient to prepare.

The goal of the canning process is to destroy any microorganisms in the food and prevent recontamination by microorganisms. Heat is the most common agent used to destroy microorganisms. Removal of oxygen can be used in conjunction with other methods to prevent the growth of oxygen-requiring microorganisms.

In the conventional canning of fruits and vegetables, there are basic process steps that are similar for both types of products. However, there is a great diversity among all plants and even those plants processing the same commodity. The differences include the inclusion of certain operations for some fruits or vegetables, the sequence of the process steps used in the operations, and the cooking or blanching steps. Production of fruit or vegetable juices occurs by a different sequence of operations and there is a wide diversity among these plants. Typical canned products include beans (cut and whole), beets, carrots, corn, peas, spinach, tomatoes, apples, peaches, pineapple, pears, apricots, and cranberries. Typical juices are orange, pineapple, grapefruit, tomato, and cranberry. Generic process flow diagrams for the canning of fruits, vegetables, and fruit juices are shown in Figures 5.1, 5.2, and 5.3. The steps outlined in these figures are intended to represent the basic processes in production. A typical commercial canning operation may employ the following general processes: washing, sorting/grading, preparation, container filling, exhausting, container sealing, heat sterilization, cooling, labeling/casing, and storage for shipment. In these diagrams, no attempt has been made to be product specific and include all process steps that would be used for all products.

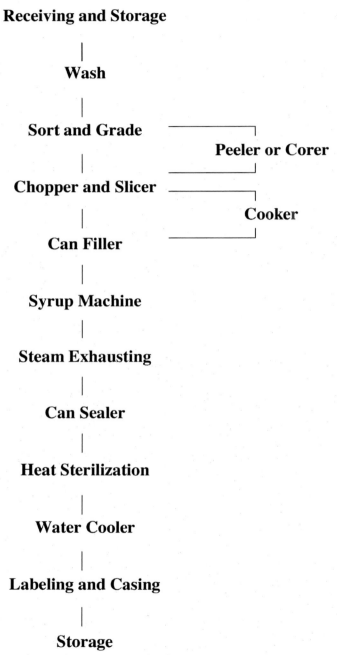

FIGURE 5.1
Generic process flow diagram for fruit canning.

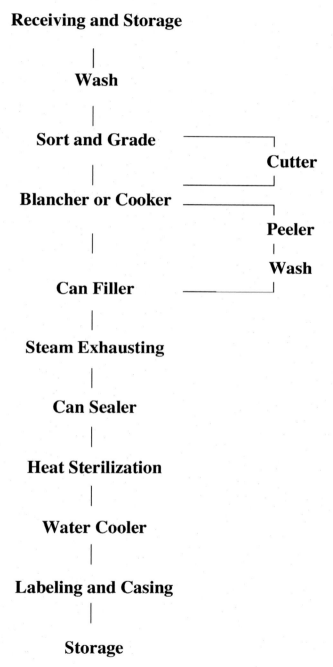

FIGURE 5.2
Generic process flow diagram for vegetable canning.

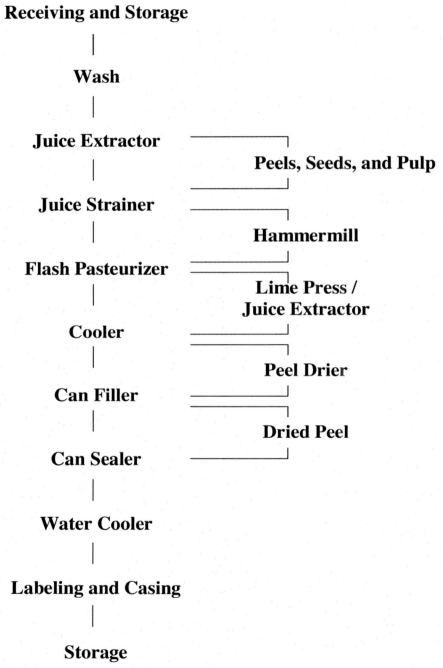

FIGURE 5.3
Generic process flow diagram for juice canning.

One of the major differences in the sequence of operations between fruit and vegetable canning is the blanching operation. Most of the fruits are not blanched prior to can filling, whereas many of the vegetables undergo this step. Canned vegetables generally require more severe processing than do fruits because the vegetables have much lower acidity and contain more heat-resistant soil organisms. Many vegetables also require more cooking than fruits to develop their most desirable flavor and texture. The methods used in the cooking step vary widely among facilities. With many fruits, preliminary treatment steps (e.g., peeling, coring, halving, pitting) occur prior to any heating or cooking step, but with vegetables, these treatment steps often occur after the vegetable has been blanched. For both fruits and vegetables, peeling is done either by a mechanical peeler, steam peeling, or lye peeling. The choice depends upon the type of fruit or vegetable or the choice of the company (EPA 1995a).

## DID YOU KNOW?

Blanching is a cooking term that describes a process of food preparation wherein the food substance, usually a fruit or vegetable, is plunged into boiling water, removed after a brief, timed interval and finally plunged into iced water or placed under cold running water (shocked) to halt the cooking process.

Some citrus fruit processors produce dry citrus peel, citrus molasses, and d-Limonene from the peels and pulp residue collected from the canning and juice operations. Other juice processing facilities use concentrates, and raw commodity processing does not occur at the facility. The peels and residue are collected and ground in a hammermill, lime is added to neutralize the acids, and the product pressed to remove excess moisture. The liquid from the press is screened to remove large particles, which are recycled back to the press, and the liquid is concentrated to molasses in an evaporator. The pressed peel is sent to a direct-fired hot-air drier. After passing through a condenser to remove the d-Limonene, the exhaust gases from the drier are used as the heat source for the molasses evaporator (EPA 1995a).

Equipment for conventional canning has been converting from batch to continuous units. In continuous retorts, the cans are fed through an air lock, then rotated through the pressurized heating chamber, and subsequently cooled through a

## WHAT IS D-LIMONENE?

The major component of the oil extracted from citrus rind is d-Limonene. When citrus fruits are juiced, the oil is pressed out of the rind. This oil is separated from the juice and distilled to recover certain flavor and fragrance compounds. The bulk of the oil is left behind and collected. This is food grade d-Limonene. After the juicing process, the peels are conveyed to a steam extractor. This extracts more of the oil from the peel. When the steam is condensed, a layer of oil floats on the surface of the condensed water. This is technical grade d-Limonene. (Florida Chemical Company 2007)

second section of the retort in a separate cold-water cooler. Commercial methods for sterilization of canned foods with a pH of 4.5 or lower include use of static retorts, which are similar to large pressure cookers. A newer unit is the agitating retort, which mechanically moves the can and the food, providing quicker heat penetration. In the aseptic packaging process, the problem with slow heat penetration in the in-container process is avoided by sterilizing and cooling the food separate from the container. Presterilized containers are then filled with the sterilized and cooled product and are sealed in a sterile atmosphere (EPA 1995a).

### EXAMPLE PROCESS: CANNING WHOLE TOMATOES

To provide a closer insight into the actual processes (and potential hazard exposures) that occur during a canning operation, the EPA's eight-step description of the canning of whole tomatoes is presented in the following paragraphs. This description provides more detail for each of the operations than is presented in the generic process flow diagrams in Figures 5.1, 5.2, and 5.3.

#### Step 1—Preparation

The principal preparation steps are washing and sorting. Mechanically harvested tomatoes are usually thoroughly washed by high-pressure sprays or by strong-flowing streams of water while being passed along a moving belt or on agitating or revolving screens. The raw produce may need to be sorted for size and maturity. Sorting for size is accomplished by passing the raw tomatoes through a series of moving screens with different mesh sizes or over differently spaced rollers. Separation into groups according to degree of ripeness or perfection of shape is done by hand; trimming is also done by hand.

### Step 2—Peeling and Coring

Formerly, tomatoes were initially scalded followed by hand peeling, but steam peeling and lye peeling have also become widely used. With steam peeling, the tomatoes are treated with steam to loosen the skin, which is then removed by mechanical means. In lye peeling, the fruit is immersed in a hot lye bath or sprayed with a boiling solution of 10 to 20 percent lye. The excess lye is then drained and any lye that adheres to the tomatoes is removed with the peel by thorough washing.

Coring is done by a water-powered device with a small turbine wheel. A special blade mounted on the turbine wheel spins and removes the tomato cores.

### Step 3—Filling

After peeling and coring, the tomatoes are conveyed by automatic runways, through washers, to the point of filling. Before being filled, the cans or glass containers are cleaned by hot water, steam, or air blast. Most filling is done by machine. The containers are filled with the solid product and then usually topped with a light puree of tomato juice. Acidification of canned whole tomatoes with 0.1 to 0.2 percent citric acid has been suggested as a means of increasing acidity to a safer and more desirable level. Because of the increased sourness of the acidified product, the addition of 2 to 3 percent sucrose is used to balance the taste. The addition of salt is important for palatability.

### Step 4—Exhausting

The objective of exhausting containers is to remove air so that the pressure inside the container following heat treatment and cooling will be less than atmospheric. The reduced internal pressure (vacuum) helps to keep the can ends drawn in, reduces strain on the containers during processing, and minimizes the level of oxygen remaining in the headspace. It also helps to extend the shelf life of food products and prevents bulging of the container at high altitudes.

Vacuum in the can may be obtained by the use of heat or by mechanical means. The tomatoes may be preheated before filling and sealed hot. For products that cannot be preheated before filling, it may be necessary to pass the filled containers through a steam chamber or tunnel prior to the sealing machine to expel gases from the food and raise the temperature. Vacuum also may be produced mechanically by sealing containers in a chamber under a high vacuum.

### Step 5—Sealing

In sealing lids on metal cans, a double seam is created by interlocking the curl of the lid and flange of the can. Many closing machines are equipped to create vacuum in the headspace either mechanically or by steam-flow before lids are sealed.

### Step 6—Heat Sterilization

During processing, microorganisms that can cause spoilage are destroyed by heat. The temperature and processing time vary with the nature of the product and the size of the container.

Acidic products, such as tomatoes, are readily preserved at 100°C (212°F). The containers holding these products are processed in atmospheric steam or hot-water cookers. The rotary continuous cookers, which operate at 100°C (212°F), have largely replaced retorts and open-still cookers for processing canned tomatoes. Some plants use hydrostatic cookers and others use continuous-pressure cookers.

### Step 7—Cooling

After heat sterilization, containers are quickly cooled to prevent overcooking. Containers may be quick-cooled by adding water to the cooker under air pressure or by conveying the containers from the cooker to a rotary cooler equipped with a cold-water spray.

### Step 8—Labeling and Casing

After the heat sterilization, cooling, and drying operations, the containers are ready for labeling. Labeling machines apply glue and labels in one high-speed operation. The labeled cans or jars are then packed into shipping cartons (EPA 1995a).

## DEHYDRATED FRUITS AND VEGETABLES

Dehydration of fruit and vegetables is one of the oldest forms of food preservation techniques known to man and consists primarily of establishments engaged in sun-drying or artificially dehydrating fruits and vegetables. Although food preservation is the primary reason for dehydration, dehydration of fruits and vegetables also lowers the cost of packaging, storing, and transportation by reducing both the weight and volume of the final product. (The reduction in bulk and weight is particularly attractive to campers and backpackers.) Given the improvement in the quality of dehydrated foods, along with the increased focus on instant and convenience foods, the potential for increased production of dehydrated fruits and vegetables is greater than ever (EPA 1995b).

In 2002, there were 147 companies involved in dried and dehydrated food manufacturing. These companies produced over 5 billion dollars in inventory and employed approximately 15,400 people. These figures are an increase from 1997, when 125 companies employed 14,600 people. California was the largest producer, accounting for 37 percent of the nation's dried and dehydrated food manufacturing. Other major producers (in descending order) were Idaho, Illinois, Texas, New York, and Michigan (DOC 2004).

According to Somogyi and Luh (1986, 1988) and the U.S. Department of Commerce (2004), "dehydrated" fruits and vegetables are generally defined as food that has had its moisture content reduced to a level below which microorganisms can grow (8 to 18 percent moisture). "Dried" fruit and vegetables are defined as food that has had a reduction in moisture content in general, and has moisture content below 30 percent. The moisture content of most dehydrated food is below 20 percent, and depends on the drying process used. Intermediate-moisture or semi-moist foods contain 15 to 30 percent moisture. Most fruit is dried using sun (solar) drying, while most vegetables are dried using continuous forced-air processes.

### Process Description

Dried or dehydrated fruits and vegetables can be produced by a variety of processes. These processes differ primarily by the type of drying method used, which depends on the type of food and the type of characteristics of the final product. In general, dried or dehydrated fruits and vegetables undergo the following process steps: predrying treatments, such as size selection, peeling, and color preservation; drying or dehydration, using natural or artificial methods; and postdehydration treatments, such as sweating, inspection, and packaging (EPA 1995b).

Figure 5.4 shows a process flow diagram for a typical fruit or vegetable dehydration process. In general, "dried" refers to all dried products, regardless of the method of drying, and "dehydrated" refers to products that use mechanical equipment and artificial heating methods (as opposed to natural drying methods) to dry the product (EPA 1995a).

## WORKER SAFETY AND HEALTH CONCERNS

Based on the fruit and vegetable processes and dehydration operations described above, the reader should be able to discern many of the potential hazards workers are (or can be) exposed to during each workshift. During each of the processes described workers can be exposed to, to name a few, machines or mechanical transmission apparatus power systems; many forms of energy and stored energy (kinetic and potential energy sources); electrical wiring and components; process chemicals (e.g., lime) and off-gases generated from the processes; wet, hot, cold, and/or humid work areas; slip, trip, and fall hazards; and a wide assortment of powered industrial trucks.

As is the case with other segments of the food processing industry, the regulations governing labor practices for the fruit and vegetable processing industry are listed under Title 29 of the Code of Federal Regulations. The specific occupational safety and

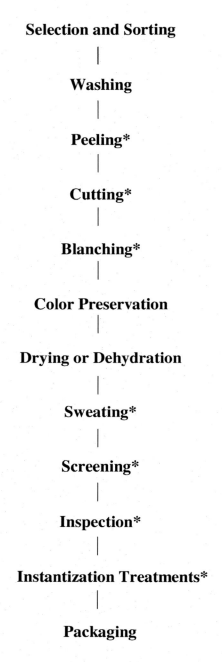

**Selection and Sorting**

|

**Washing**

|

**Peeling***

|

**Cutting***

|

**Blanching***

|

**Color Preservation**

|

**Drying or Dehydration**

|

**Sweating***

|

**Screening***

|

**Inspection***

|

**Instantization Treatments***

|

**Packaging**

*May not occur for all fruits and vegetables.*
*Source: U.S. EPA (1995)*

FIGURE 5.4
Generic process flow diagram for dehydration of fruits and vegetables.

health regulations we are concerned with are found in 29 CFR Part 1910 (General Industry Standards), Part 1926 (Construction Standards), and Part 1928 (Agriculture).

As with the meatpacking industry, there are currently no specific OSHA standards for
the fruit and vegetable processing industry. However, in the following we highlight those
general industry standards related to or applicable to fruit and vegetable processing:

### General Industry (29 CFR 1910)
- 1910 Subpart D, Walking-working surfaces
  - 1910.21, Definitions
  - 1910.22, General requirements
  - 1910.23, Guarding floor and wall openings and holes
  - 1910.24, Fixed industrial stairs
  - 1910.25, Portable wood ladders
  - 1910.26, Portable metal ladders
  - 1910.27, Fixed ladders
- 1910 Subpart G, Occupational health and environmental control
  - 1910.95, Occupational noise exposure
- 1910 Subpart H, Hazardous materials
  - 1910.119, Process safety management of highly hazardous chemicals
- 1910 Subpart I, Personal protective equipment
- 1910 Subpart J, General environmental controls
  - 1910.147, The control of hazardous energy (lockout/tagout)
- 1910 Subpart O, Machinery and machine guarding
  - 1910.212, General requirements for all machines
    - 1910.212(a)(3)(ii)
- 1910 Subpart P, Hand and portable powered tools and other handheld equipment
  - 1910.243, Guarding of portable powered tools
- 1910 Subpart S, Electrical
- 1910 Subpart Z, Toxic and hazardous substances
  - 1910.1200, Hazard communication

Because of the potential for employee exposure to hazards, it is not surprising that
OSHA keeps a close eye on the fruit and vegetable processing industry. In addition to
monitoring the industry, OSHA typically cites processing plants for various violations
of standards. Consider, for example, the top 10 violations cited by OSHA for the fruit
and vegetable industry for the 2005 fiscal year (see table 5.1).

In addition to typical physical and chemical hazards that fruit and vegetable process
workers may be exposed to during a shift, air emissions (requiring ventilation protection) may also arise from a variety of sources in the canning/dehydration of fruits and

**Table 5.1.   Top 10 violations cited (FY 2005)**

| Standard | # Cited | # Inspected | Description |
|---|---|---|---|
| 1910.219 | 53 | 25 | Mechanical Power—Transmission Apparatus |
| 1910.147 | 52 | 29 | The Control of Hazardous Energy, Lockout/Tagout |
| 1910.212 | 44 | 32 | Machines, General Requirements |
| 1910.305 | 35 | 23 | Electrical, Wiring Methods, Components and Equipment |
| 1910.1200 | 32 | 19 | Hazard Communication |
| 1910.23 | 31 | 23 | Guarding Floor and Wall Openings and Holes |
| 1910.146 | 31 | 11 | Permit-Required Confined Spaces |
| 1910.303 | 30 | 22 | Electrical Systems Design, General Requirements |
| 1910.178 | 29 | 21 | Powered Industrial Trucks |
| 1910.132 | 22 | 15 | Personal Protective Equipment, General Requirements |

Source: OSHA IMIS Database (2005).

vegetables. Along with being potential hazardous exposures for workers, these emissions may also contaminate the environment (Jones et al. 1979; Woodroof and Luh 1986; Van Langenhove et al. 1991; Buttery et al. 1990; Rafson 1977).

Particulate matter (PM) emissions result mainly from solids handling, solids size reduction, and drying (e.g., citrus peel driers). Some of the particles are dusts, but others (particularly those from thermal processing operations) are produced by condensation of vapors and may be in the low-micrometer or submicrometer particle-size range.

Volatile organic compound (VOC) emissions may potentially occur at almost any stage of processing, but most usually are associated with thermal processing steps, such as cooking, and evaporative concentration. The cooking technologies in canning processes are very high-moisture processes, so the predominant emissions will be steam or water vapor. The waste gases from these operations may contain particulate matter or, perhaps, condensable vapors, as well as malodorous VOCs. Particulate matter, condensable materials, and the high moisture content of the emissions may interfere with the collection or destruction of these VOCs. The condensable materials also may be malodorous (EPA 1995a).

## DID YOU KNOW?

VOCs are organic chemicals that have a high vapor pressure and easily form vapors at normal temperature and pressure.

## REFERENCES AND RECOMMENDED READING

Buttery, R .G., et al. 1990. Identification of Additional Tomato Paste Volatiles. *Journal of Agricultural and Food Chemistry* 38 (3): 792–795.

Deroiser, N. W. 1970. *The Technology of Food Preservation.* 3rd ed. Westport, CT: Avi.

Florida Chemical Company. 2007. *Citrus oils and d-Limonene.* Winter Haven, FL: Florida Chemical Company.

Jones, J. L., et al. 1979. Overview of Environmental Control Measures and Problems in the Food Processing Industries. Industrial Environmental Research Laboratory, Cincinnati, OH, Kenneth Dostal, Food and Wood Products Branch. Grant No. R804642-01.

Luh, B. S., and J. G. Woodroof, eds. 1988. *Commercial Vegetable Processing.* 2nd ed. New York: Van Nostrand Reinhold.

Occupational Safety and Health Administration (OSHA). 2005. Top 10 Violations Cited. www.osha.gov/dep/industry_profiles/p_profile-203.html#section5 (accessed August 16, 2007).

———. 2007. Preserved Fruits and Vegetables. www.osha.gov/dep/industry_profiles/ p_profile-203.html (accessed August 13, 2007).

Rafson, H. J. 1977. Odor Emission Control for the Food Industry. *Food Technology,* June.

Somogyi, L. P., and B. S. Luh. 1986. Dehydration of Fruits, Commercial Fruit Processing. In *Commercial Fruit Processing,* 2nd ed. Ed. by J. G. Woodroof and B. S. Luh. Westport, CT: Avi.

———. 1988. Vegetable Dehydration, Commercial Vegetable Processing. In *Commercial Fruit Processing,* 2nd ed. Ed. by J. G. Woodroof and B. S. Luh. Westport, CT: Avi.

U.S. Department of Commerce (DOC). 2004. *Dried and Dehydrated Food Manufacturing: 2002.* 2002 Economic Census. Manufacturing Industry Series. Washington, DC: U.S. Department of Commerce, U.S. Census Bureau. www.census.gov/prod/ec02/ ec0231i311423.pdf (accessed March 21, 2008).

U.S. Environmental Protection Agency (EPA). 1995a. Canned Fruits and Vegetables. www.epa.gov/ttn/chief/ap42/ch09/final/c9s08-1.pdf (accessed August 14, 2007).

———. 1995b. Dehydrated Fruits and Vegetables. www.epa.gov/ttn/chief/ap42/ch09/ final/c9s08-2.pdf (accessed August 14, 2007).

Van Langenhove, H. J., et al. 1991. Identification of Volatiles Emitted during the Blanching Process of Brussels Sprouts and Cauliflower. *Journal of the Science of Food and Agriculture* 55: 483–487.

Woodroof, J. G., and B. S. Luh, eds. 1986. *Commercial Fruit Processing.* 2nd ed. Westport, CT: Avi.

# The Food Flavorings Industry and Popcorn Workers' Lung

### Popcorn Production Harms Workers

One of America's favorite snack foods—popcorn—is at the center of a national health controversy. The chemical diacetyl, used to make artificial butter flavoring, has been linked to a respiratory disease called "popcorn lung" [bronchiolitis obliterans] in hundreds of people. Labor unions and prominent occupational health scientists are calling on federal authorities to set an emergency standard for the chemical in the workplace.

—Living on Earth, 2006

### The Federal Response?

110th Congress—1st Session

### H.R. 2693

To direct the Occupational Safety and Health Administration to issue a standard regulating worker exposure to diacetyl.

### A BILL

*Be it enacted by the Senate and House of Representatives of the United States of America in Congress assembled,*

### SECTION 1. FINDINGS.

Congress finds the following:

(1) An emergency exists concerning worker exposure to diacetyl, a substance used in many flavorings, including artificial butter flavorings.

(2) There is compelling evidence that diacetyl presents a grave danger and significant risk of life-threatening illness to exposed employees. Workers exposed to diacetyl have developed, among other conditions, a debilitating lung disease known as bronchiolitis obliterans.

(3) From 2000–2002 NIOSH identified cases of bronchiolitis obliterans in workers employed in microwave popcorn plants, and linked these illnesses to exposure to diacetyl used in butter flavoring. In December 2003, NIOSH issued an alert "Preventing Lung Disease in Workers Who Use or Make Flavorings," recommending that employers implement measures to minimize worker exposure to diacetyl.

(4) In August 2004 the Flavor and Extract Manufacturers Association (FEMA) of the United States issued a report, "Respiratory Health and Safety in the Flavor Manufacturing Workplace," warning about potential serious respiratory illness in workers exposed to flavorings and recommending comprehensive control measures for diacetyl and other "high priority" substances used in flavoring manufacturing.

(5) From 2004–2007 additional cases of bronchiolitis obliterans were identified among workers in the flavoring manufacturing industry by the California Department of Health Services and Division of Occupational Safety and Health (Cal/OSHA), which through enforcement actions and an intervention program called for the flavoring manufacturing industry in California to reduce exposure to diacetyl.

(6) In a report issued in April 2007, NIOSH reported that flavor manufacturers and flavored-food producers are widely distributed in the United States and that bronchiolitis obliterans had been identified among microwave popcorn and flavoring-manufacturing workers in a number of states.

(7) Despite NIOSH's findings of the hazards of diacetyl and recommendations that exposures be controlled, and a formal petition by labor organizations and leading scientists for issuance of an emergency temporary standard, the Occupational Safety and Health Administration (OSHA) has not acted to promulgate an occupational safety and health standard to protect workers from harmful exposure to diacetyl.

(8) An OSHA standard is urgently needed to protect workers exposed to diacetyl from bronchiolitis obliterans and other debilitating conditions.

## FLAVORING SUBSTANCES AND COMPOUNDED FLAVORS

A flavor to be added to food is a complex mixture of individual flavoring substances that has been "compounded" to provide the desired taste perception, or "flavor." For example, the flavor humans perceive as "orange" is a complex mixture of over 100 individual substances that each contribute to the flavor as it is perceived.

There are over 2,000 individual single chemically defined flavoring substances used by flavor manufacturers to formulate flavors. Natural extracts (e.g., vanilla extract) are also used to formulate flavors, and may be directly added to foods in the manufacturing process. Compounded flavors and extracts usually contain a solvent such as propylene glycol or triacetin to facilitate the use of the flavor.

The safety of flavors when they are added to foods has been thoroughly evaluated by the FEMA Expert Panel (e.g., Smith et al. 2003) and flavors are strictly regulated under their conditions of intended use in foods by the Food and Drug Administration (e.g., 21 CFR Parts 172, 182) (FEMA 2004).

## WORKERS WHO USE OR MAKE FLAVORINGS

The occurrence of severe lung disease in workers who make flavorings or use them to produce microwave popcorn has revealed an unrecognized occupational health risk. Flavorings are often complex mixtures of many chemicals (Conning 2000).

WARNING!

Breathing certain flavoring chemicals in the workplace may lead to severe lung disease. (CDC 2005)

The safety of these chemicals is usually established for humans consuming small amounts in foods (Pollitt 2000), not for food industry workers inhaling them. Production workers employed by flavoring manufacturers (or those who use flavorings in the production process) often handle a large number of chemicals, many of which can be highly irritating to breathe in high concentrations.

This Alert describes health effects that may occur because of workplace exposure to some flavorings or their ingredients, gives examples of workplace settings in which illness has occurred, and recommends steps that companies and workers should take to prevent hazardous exposures.

## Background

NIOSH has investigated the occurrence of severe lung disease in workers at a microwave popcorn packaging plant. Eight former workers at this plant developed illness characterized by fixed airways obstruction on lung function tests (Akpinar-Elci et al. 2002). An evaluation of the current workforce at this plant showed an association between exposure in vapors from flavorings used in the production process and decreased lung function (Kreiss, Gomaa, Kullman et al. 2002). Similar fixed obstructive lung disease has also occurred in workers at other plants that use or manufacture flavorings (NIOSH 1986; Lockey et al. 2002). In animal tests, inhaling vapors from a heated butter flavoring used in microwave popcorn production caused severe injury to airways (Hubbs, Battelli, Goldsmith et al. 2002).

Medical test results in affected workers (including some lung biopsy results) are consistent with bronchiolitis obliterans, an uncommon lung disease characterized by fixed airways obstruction (Akpinar-Elci et al. 2002). In bronchiolitis obliterans, inflammation and scarring occur in the smallest airways of the lung and can lead to severe and disabling shortness of breath. The disease has many known causes such as inhalation of certain chemicals, certain bacterial and viral infections, organ transplantation, and reactions to certain medications (King 2000). Known causes of bronchiolitis obliterans due to occupational or other environmental exposures include gases such as nitrogen oxides (e.g., silo gas), sulfur dioxide, chlorine, ammonia, phosgene, and other irritant gases (King 1998). Recent NIOSH investigations

## SILO GASES

Silo gas is formed by the natural fermentation of chopped silage shortly after it is placed in the silo. The nitrogen oxides formed are harmful because they cause severe irritation to the nose and throat and may lead to inflammation of the lungs. However, what makes this gas especially dangerous is that lower level exposure to it is accompanied by only a little immediate pain or discomfort, though death can and has occurred immediately after contact with high concentrations. A farmer might breathe the gas without noticing any serious ill effects and then die in his sleep hours later from fluid collecting in his lungs.

strongly suggest that some flavoring chemicals can also cause bronchiolitis obliterans in the workplace. (Some workers exposed to flavorings in one of these plants were also found to have occupational asthma.)

**Health Effects**

The main respiratory symptoms experienced by workers affected by fixed airways obstruction include cough (usually without phlegm) and shortness of breath on exertion. These symptoms typically do not improve when the worker goes home at the end of the workday or on weekends or vacations. The severity of the lung symptoms can range from only a mild cough to severe cough and shortness of breath on exertion. Usually these symptoms are gradual in onset and progressive, but severe symptoms can occur suddenly. Some workers may experience fever, night sweats, and weight loss. Before arriving at a final diagnosis, doctors of affected workers initially thought that the symptoms might be due to asthma, chronic bronchitis, emphysema, pneumonia, or smoking. Severe cases may not respond to medical treatment. Affected workers generally notice a gradual reduction or cessation of cough years after they are no longer exposed to flavoring vapors, but shortness of breath on exertion persists. Several with very severe disease were placed on lung transplant waiting lists. Workers exposed to flavorings may also experience eye, nose, throat, and skin irritation. In some cases, chemical eye burns have required medical treatment.

*Medical Evaluation*
Medical testing may reveal several of the following findings:

- Spirometry, a type of breathing test,
  - most often shows fixed airways obstruction (i.e., difficulty blowing air out fast and no improvement with asthma medications)
  - sometimes shows restriction (i.e., decreased ability to fully expand the lungs)
- Lung volumes may show hyperinflation (i.e., too much air in the lungs due to air trapping beyond obstructed airways).
- Diffusing capacity of the lung (DLCO) is generally normal, especially early in the disease.
- Chest X-rays are usually normal but may show some hyperinflation.
- High-resolution computerized tomography scans of the chest at full inspiration and expiration may reveal heterogeneous air trapping on the expiratory view as well as haziness and thickened airway walls.
- Lung biopsies may reveal evidence of constrictive bronchiolitis obliterans (i.e., severe narrowing or complete obstruction of the small airways). An open lung biopsy, such as by thoracoscopy, is more likely to be diagnostic than a transbronchial biopsy. Special processing, staining, and review of multiple tissue sections may be necessary for a diagnosis.

### Current Exposure Limits

Flavorings are composed of various natural and manmade substances. They may consist of a single substance, but more often they are complex mixtures of several substances. The Flavor and Extract Manufacturers Association (FEMA) evaluates flavoring ingredients to determine whether they are "generally recognized as safe" (GRAS) under the conditions of intended use through food consumption. Though considered safe to eat, ingredients may be harmful to breathe in the forms and concentrations to which food and chemical industry workers may be exposed.

Occupational exposure guidelines have been developed for only a small number of the thousands of ingredients used in flavorings. For example, Occupational Safety and Health Administration (OSHA) permissible exposure limits (PELs) and/or NIOSH recommended exposure limits (RELs) have been established for only 46 (<5%) of the 1,037 flavoring ingredients considered by the flavorings industry to represent potential respiratory hazards due to possible volatility and irritants properties (alpha, beta-unsaturated aldehydes and ketones, aliphatic aldehydes, aliphatic carboxylic acids, aliphatic amines, and aliphatic aromatic thiols and sulfides) (Hallagan 2002). Material safety data sheets (MSDSs) contain information about known occupational hazards of specific chemicals, but they may not be based on the most up-to-date information in the case of newly recognized occupational health risks (CDC 2005).

## CASE STUDY 1

In September 2003, a man aged 29 years with no history of smoking, lung disease, or respiratory symptoms developed progressive shortness of breath on exertion, decreased exercise tolerance, intermittent wheezing, left-sided chest pain, and a productive cough two years after beginning employment as a flavor compounder. His job involved measuring diacetyl and other ingredients to prepare batches of powder flavorings. The workplace did not have effective methods for controlling exposure to the flavoring chemicals, such as local exhaust ventilation or adequate use of respirators to reduce exposure to organic compounds and powders. The worker reported wearing a paper dust mask and occasionally a cartridge respirator for organic vapors. However, he never received a fit test for the respirator. He had a beard at the time, which precluded a proper fit, and he was not adequately protected from both volatile organic chemicals and particulates.

In November 2003, the man went to his primary-care physician and was treated with antibiotics and bronchodilators for suspected bronchitis and allergic rhinitis. In January 2004, he stopped working because of his respiratory symptoms. His shortness of breath became more severe, with dyspnea after walking 10–15 feet. A high-resolution computed tomography (HRCT) scan of his chest showed cylindrical bronchiectasis in the lower lobes, with scattered peribronchial ground-glass opacities. In April 2004, spirometry showed severe obstructive lung disease, with a forced expiratory volume in one second (FEV1) of 28 percent of the predicted normal value, without bronchodilator response. Static lung volumes by body plethysmography were consistent with severe air trapping. Diffusing capacity was normal.

In October 2004, the patient was referred for an occupational pulmonary consultation. Paired inspiratory and expiratory HRCT scans showed central peribronchial thickening with central airway dilatation and subtle areas of mosaic attenuation scattered throughout the lungs, predominantly in the right lower lobe. The diagnosis of work-related bronchiolitis obliterans was made on the basis of history, fixed airway obstruction with normal diffusing capacity, and typical HRCT findings (California Department of Health Services 2006). Diacetyl is considered the cause of this patient's disease on the basis of its known toxic effects; however, exposure to other less well-characterized flavoring chemicals might also have contributed (CDC 2007).

CASE STUDY

## CASE STUDY 2

During 2002, a nonsmoking woman aged 40 years, who had no history of lung disease or respiratory symptoms when she began working as a flavor compounder, experienced nasal congestion and cough after five years on the job, which involved mixing dry powders with diacetyl and other ingredients to make artificial butter flavoring. The workplace did not have exposure-control measures such as local exhaust ventilation, and employees did not use respirators appropriately. The worker reported wearing a paper dust mask that had not been fit tested and did not provide adequate protection from either volatile organic compounds or particulates. The woman was treated with antibiotics and antihistamines by her primary-care physician. She experienced progressively worsening shortness of breath on exertion, decreasing exercise tolerance, and a nonproductive cough. In November 2005, she visited a pulmonary specialist who suspected work-related asthma and treated her with bronchodilators and oral corticosteroids, producing minimal improvement. An HRCT of the chest showed several small areas of patchy ground-glass opacities throughout the lungs.

In December 2005, the patient stopped working because of her respiratory symptoms. Spirometry revealed severe obstructive lung disease, with an FEV1 of 18 percent of the predicted normal value, without bronchodilator response. Static lung volumes by body plethysmography were consistent with severe air trapping. Diffusing capacity was normal. Left thoracotomy with wedge resection of the left lower lobe did not indicate bronchiolitis obliterans in this area of the lung. However, other findings of peribronchial inflammation, interstitial fibrosis, and non-caseating-type granulomas suggested an inflammatory process. The diagnosis of work-related bronchiolitis obliterans was made on the basis of history, fixed airway obstruction with normal diffusing capacity, and typical HRCT findings (California Department of Health Services 2006) (CDC 2007).

## OSHA COMPLIANCE STANDARDS: FLAVORINGS INDUSTRY

There are currently no specific standards for flavorings-related lung disease. However, OSHA standards regulating all workplaces offer protection to workers exposed to these substances.

In addition to the applicable compliance requirements under the general duty clause, the OSHA General Industry standards that apply to the flavorings industry are listed below.

*Highlighted Standards*
**General Industry (29 CFR 1910)**

- 1910 Subpart I, Personal protective equipment (PPE)
  - 1910.132, General requirements
  - 1910.133, Eye and face protection
  - 1910.134, Respiratory protection
    - Appendix A, Fit testing procedures
    - Appendix B-1, User seal check procedures
    - Appendix B-2, respiratory cleaning protection procedures
    - Appendix C, OSHA respirator medical evaluation form
    - Appendix D, Information for employees using respirators
  - 1910.138, Hand protection
- 1910 Subpart Z, Toxic and hazardous substances
  - 1910.1000, Air contaminants
    - Table Z-1, Limit for air contaminants
  - 1910.1200, Hazard communication

## SAFE WORK PRACTICES FOR THE FLAVORINGS INDUSTRY

The following recommendations are provided to reduce hazardous exposures associated with the use or manufacture of flavorings. In general, NIOSH recommends that employers and workers implement controls to limit worker exposure. In order of preference, the major types of controls include the following:

1. Substitution
2. Engineering controls
3. Administrative controls
4. Education
5. Personal protective equipment
6. Exposure and worker health monitoring

### Substitution

Substituting a less hazardous material can effectively reduce an existing hazard. However, substitution does not always represent a feasible or definitive approach. An adequate substitute may not exist; or, as with flavoring mixtures, the exposures may be complex and toxicities may be inadequately understood. Therefore, do the following when considering substitution:

- Exercise extreme care when selecting substitutes.
- Consider the possible adverse health effects of any candidate substitutes.
- Remember that as a general rule, flavoring formulations designed to release less volatile chemicals or respirable powder into the air during handling may pose less risk to workers.

### Engineering Controls

Engineering controls are the primary methods for minimizing exposure associated with the use or manufacture of potentially hazardous flavorings. Examples include closed production systems (e.g., to eliminate handling open containers of flavorings or their chemical ingredients for placement into mixing tanks), adequate ventilation, and isolation.

- Whenever possible, use closed processes to transfer flavorings or their chemical ingredients.
- Isolate the mixing room and other areas where flavorings and their ingredients are openly handled. Maintain these work areas under negative air pressure relative to the rest of the plant.
- Use local exhaust ventilation of tanks and other sources of potential exposure (e.g., places where flavorings are openly weighed or measured) as well as general dilution ventilation of the work area to eliminate or reduce possible worker exposures. Obtain information about the design of appropriate ventilation systems from a qualified ventilation engineer or from *Industrial Ventilation—A Manual of Recommended Practice for Design* (ACGIH 2007).
- Check ventilation equipment regularly for adequate performance, especially in areas where flavorings and their ingredients are handled (e.g., mixing room) and in adjacent work areas. Also perform checks whenever a process change is made or a problem is suspected.
- For processes involving heating of flavorings, keep the temperature as low as possible to minimize emissions of volatile chemicals into the air.

**Administrative Controls**

- Establish and enforce work practices to limit release of chemicals and dust into the workplace air when flavorings or their ingredients are handled.
- Tightly seal containers with unused or residual amounts of flavorings or their ingredients.
- Maintain good general housekeeping in any areas where flavorings or their ingredients are handled.
- Establish standard procedures for cleaning the workplace, tanks and other containers, and spills.
  - Do not use compressed air for cleaning powdered flavorings or ingredients, as this will increase concentrations of airborne particulate.
  - Use special caution when removing residual chemicals from tanks and other containers with steam or hot water, as this may increase exposure to volatile chemical vapors.
  - Clean up spills of flavorings or their ingredients promptly using procedures and appropriate protective equipment designed to limit exposure.
- Restrict access to all areas where flavorings are openly handled; only essential workers should enter these areas and only when properly protected (see section on personal protective equipment, below).

**Employer and Worker Education (Training)**

Employer awareness of hazardous exposures in the production process and communication of this information to workers are vital elements in an optimal occupational safety and health program.

- Inform workers about any materials that may contain flavoring agents and tell them the nature of the hazard.
- Provide general information and specific hazard warnings through workplace postings, container labeling, MSDSs, and training.
- Train workers regarding the means available at the facility to eliminate or limit exposure and how they can take action to limit potential exposures for themselves and fellow workers.
- Inform workers about symptoms that may indicate a flavoring-related health problem. Advise them to report these symptoms to their supervisors and physicians.

**Personal Protective Equipment**

Whenever the substances and amounts present in a plant or work area pose a potential hazard, provide personal protective equipment to protect workers from skin, eye, and respiratory tract irritation and other adverse health effects.

*Skin and eye protection*

- Enforce the use of chemical-resistant gloves and tight-fitting goggles for workers with potential skin and eye exposure to irritant flavorings or their chemical ingredients.
- Establish specific guidance about when to use the equipment for each job, based on knowledge of the tasks performed, substances involved, and an assessment of potential exposures.

*Respiratory Protection (OSHA 29 CFR 1910.134)*

The use of respirators is the ***least preferred method*** of controlling worker exposures to respiratory hazards.

- Do not use respirators as the primary control for routine operations. However, they may be needed and used while optimal engineering controls and work practices are being implemented, during some short-duration maintenance procedures, and during emergencies.
- Use respirators for exposure situations in which even the lowest concentrations achievable with engineering controls are still associated with risk.
- The minimum protective respirator that should be used for workers exposed to flavorings or their chemical ingredients is a NIOSH-certified half-mask, negative-pressure respirator with organic vapor cartridges or canisters and particulate filters.
- Use a full-facepiece respirator for eye protection as well as additional respiratory protection.
- Consider other respirators for workers exposed to flavorings or their chemical ingredients: powered, air-purifying respirators (with organic vapor cartridges or canisters and particulate filters) and, for maximum respiratory protection, supplied-air respirators.
- Before using respirators, set up a written respiratory protection program that meets the requirements of the OSHA respiratory protection standard (29 CFR 1910.134).
- Designate a trained employee or supervisor to run the program and evaluate its effectiveness. Make sure that the designated person's training or experience is appropriate to the level of complexity of the program.
- Ensure that respirators selected for use are certified by NIOSH according to 42 CFR 84.
- Implement a change schedule for canisters and cartridges based on objective information or data that will ensure that canisters and cartridges are changed before the end of their service lives.

RESPIRATOR SELECTION PROCEDURE
(42 CFR PART 84-NIOSH)

Under 42 CFR Part 84, to select the correct respirator for protection against particulates (PM), the following conditions must be known:

- The identity and concentration of the particles in the workplace air.
- The OSHA or MSHA permissible exposure limit (PEL), NIOSH-recommended exposure limit, or other occupational exposure limit for the contaminant.
- The hazard ratio (HR) (i.e., the airborne particulate concentration divided by the exposure limit).
- The Assigned Protection Factor (APF) for the class of respirator (the APF should be greater than the HR).
- The immediately dangerous to life or health (IDLH) concentration, including oxygen deficiency.
- Any service life information available for combination cartridges or canisters.

Multiplying the occupational exposure limit by the APF for a respirator gives the maximum workplace concentration in which that respirator can be used. For example, if the commonly accepted APF for a half-mask respirator is 10 and the PEL is 5 milligrams per cubic meter, then 50 milligrams per cubic meter is the highest workplace concentration in which a half-mask respirator can be used against that contaminant. If the workplace concentration is greater than 50 milligrams per cubic meter, a more protective respirator (with a higher APF) should be used. In no case should an air-purifying respirator be used in IDLH atmospheres or in areas that are oxygen deficient, and you should never exceed the manufacturer's guidelines.

- Include the following in the respiratory protection program:
  - Procedures for selecting respirators
  - Medical evaluations of workers required to use respirators
  - Fit-testing procedures for tight-fitting respirators

- Procedures for proper use of respirators in routine and reasonably foreseeable emergency situations
- Procedures and schedules for cleaning, disinfecting, storing, inspecting, repairing, discarding, and otherwise maintaining respirators
- Procedures to ensure adequate quality, quantity, and flow of breathing air for atmosphere-supplying respirators
- Training of workers in the respiratory hazards to which they are potentially exposed during routine and emergency situations
- Training of workers in the proper use of respirators, including putting them on and removing them, any limitations on their use, and their maintenance
- Procedures for regularly evaluating the effectiveness of the program and worker compliance with program requirements.

### Exposure Monitoring

- Engage the services of a certified air sampling expert to identify the volatile flavoring chemicals that are present in significant amounts in the air, and to measure the air concentrations of one or more of these chemicals as indicators of exposure.
- When applicable, measure air concentrations of total respirable dust and the air concentration of any flavoring chemical with an OSHA PEL or a NIOSH REL.
- Use repeated monitoring to determine whether new engineering controls or changes in work practices are effectively reducing exposures.
- Continue routine monitoring on a regular basis to ensure the continuing effectiveness of controls.
- If monitoring indicates that exposure concentrations have increased, thoroughly investigate engineering controls to identify problems and guide remedial actions.

### Worker Health Monitoring

- Implement preplacement and regularly scheduled ascertainment of symptoms and spirometry testing of lung function for all workers with potentially hazardous exposure to flavorings or flavoring ingredients.
- Follow the latest American Thoracic Society guidelines for spirometry testing.
- Perform testing at least annually, since existing information makes it difficult to specify the interval between testing. The relatively rapid onset of severe airways obstruction in some affected workers suggests that more frequent intervals (perhaps every 3 months) may be appropriate in some situations.

## SPIROMETRY

Spirometry is a physiological test that measures how an individual inhales or exhales volumes of air as a function of time. The primary signal measured in spirometry may be volume or flow.

Spirometry is invaluable as a screening test of general respiratory health in the same way that blood pressure provides important information about general cardiovascular health. However, on its own, spirometry does not lead clinicians directly to an aetiological diagnosis. (Miller et al. 2005)

- Conduct more frequent testing if abnormalities related to flavoring exposure are detected in a particular workforce. Regardless, workers should not wait for regularly scheduled testing to report symptoms.
- Promptly refer workers for further medical evaluation if they have persistent cough; persistent shortness of breath on exertion; frequent or persistent symptoms of eye, nose, throat, or skin irritation; abnormal lung function on spirometry testing; or accelerated decline in lung function. Provide the evaluating physician with a copy of this Alert. The intent is to identify and prevent progression of work-related medical conditions. The physician should advise the worker about any suspected or confirmed medical condition that may be caused or aggravated by work exposures, about recommendations for further evaluation and treatment, and specifically about any recommended restriction of the worker's exposure (including removal from the workplace) or use of personal protective equipment. The physician should provide the employer with information about recommended restrictions of the worker's exposure or use of personal protective equipment.
- Do not rely on the absence of respiratory symptoms that occur in relation to work exposures to indicate that exposures are adequately controlled. In contrast to workers with work-related asthma, few if any workers with fixed airways obstruction from exposure to flavorings report improvement on days off work or during vacations. Also, flavoring-exposed workers who develop fixed airways obstruction may not have symptoms early in the course of their illness. Regularly scheduled spirometry is currently the best available test for early recognition of decreasing or abnormal lung function from occupational exposure to flavorings or their ingredients.

### Surveillance and Disease Reporting

- Assess the patterns of reported symptoms, abnormal spirometry, physician-advised exposure restrictions, and other available information about health effects within the workforce to identify areas, processes, and exposures that may require more intensive intervention to control exposures and prevent further adverse health effects.

## REFERENCES AND RECOMMENDED READING

Akpinar-Elci, M., R. Kanwal, and K. Kreiss. 2002. Bronchiolitis Obliterans Syndrome in Popcorn Plant Workers. *American Journal of Respiratory and Critical Care Medicine* 165: A526.

American Conference of Governmental Industrial Hygienists (ACGIH). 2007. *Industrial Ventilation: A Manual of Recommended Practice for Design.* 26th ed. Cincinnati, OH: ACGIH.

California Department of Health Services. 2006. Food Flavoring Workers with Bronchiolitis Obliterans Following Exposure to Diacetyl—California. http://www.dhs.ca.gov/ohb/flavoringcases.pdf (accessed March 24, 2008).

Centers for Disease Control and Prevention (CDC). 2005. NIOSH: Preventing Lung Disease in Workers Who Use or Make Flavorings. www.cdc.gov/niosh/docs/2004-110 (accessed August 18, 2007).

———. 2007. Fixed Obstructive Lung Disease among Workers in the Flavor-Manufacturing Industry—California, 2004–2007. *Morbidity and Mortality Weekly Report* 56 (16): 389–393.

Conning, D. M. 2000. Toxicology of Food and Food Additives. In *General and Applied Toxicology.* 2nd ed. Ed. B. Ballantyne, T. C. Marrs, and T. Syversen. London: Macmillan, 1977–1992.

Flavor and Extract Manufacturers Association of the United States (FEMA). 2004. *Respiratory Health and Safety in the Flavor Manufacturing Workplace.* Washington, DC: FEMA.

Greene, M. V. 2007. "Popcorn Workers' Lung" Spurs Tug-of-War over Regulation. *Safety + Health* 176 (2): 32–33.

Hallagan, J. B. 2002. Letter of November 26, from J. B. Hallagan, Flavor and Extract Manufacturers Association of the United States, to R. Kanwai, Division of Respiratory Disease Studies, National Institute for Occupational Safety and Health, Centers for Disease Control and Prevention, Department of Health and Human Services.

Hubbs, A. F., L. S. Battelli, W. T. Goldsmith, et al. 2002. Necrosis of Nasal and Airway Epithelium in Rates Inhaling Vapors of Artificial Butter Flavoring. *Toxicology and Applied Pharmacology* 185: 128–135.

Hubbs, A., V. Castranova, W. Jones, et al. 2002. Workplace Safety and Food Ingredients: The Example of Butter Flavoring. In Abstracts of Papers, 24th ACS National Meeting, Boston, Massachusetts, August 18–22. Washington, DC: American Chemical Society, AGFD-148.

King, T. E. 1998. Bronchiolitis. In *Pulmonary Diseases and Disorders*, ed. A. A. Fishman. New York: McGraw-Hill, 825–847.

———. 2000. Bronchiolitis. *European Respiratory Monograph* 14: 244–266.

Kreiss, K., A. Gomaa, G. Kullman, K. Fedan, E. J. Simoes, and P. L. Enright. 2002. Clinical Bronchiolitis Obliterans in Workers at a Microwave-Popcorn Plant. *New England Journal of Medicine* 347: 330–338.

Kreiss, K., A. Hubbs, and G. Kullman. 2002. Correspondence: Bronchiolitis in Popcorn-Factory Workers. *New England Journal of Medicine* 347: 1981–1982.

Living on Earth. 2006. Popcorn Production Harms Workers. www.Loe.org/shows/segments .htm? (accessed August 17, 2007).

Lockey, J., R. McKay, E. Barth, J. Dahisten, and R. Baughman. 2002. Bronchiolitis Obliterans in the Food Flavoring Manufacturing Industry. *American Journal of Respiratory and Critical Care Medicine* 165: A461.

Miller, M. R., et al. 2005. Standardisation of Spirometry. *European Respiratory Journal* 26 (2): 319–338.

National Institute for Occupational Safety and Health (NIOSH). 1986. Hazard Evaluation and Technical Assistance Report: International Bakers Services, Inc., South Bend, Indiana. NIOSH Report No. HETA 85-171-1710. Cincinnati, OH: NIOSH.

Parmet, A. J., and S. Von Essen. 2002. Rapidly Progressive, Fixed Airway Obstructive Disease in Popcorn Workers: A New Occupational Pulmonary Illness? *Journal of Occupational and Environmental Medicine* 44: 216–218.

Pollitt, F. D. 2000. Regulation of Food Additives and Food Contract Materials. In *General and Applied Toxicology*. 2nd ed. Ed. B. Ballantyne, T. C. Marrs, and T. Syversen. London: Macmillan, 1653–1660.

Smith, R. L., et al. 2003. GRAS Flavoring Substances 21. *Food Technology* 57 (5).

# Radiation Usage in the Food Industry

When people hear the word *radiation*, they generally think of nuclear power plants, nuclear weapons, or radiation treatments for cancer. As the EPA (2008c) points out, it would also be correct to add microwaves, radar, electrical power lines, cell phones, and sunshine to the list. There are many different types of radiation that have a range of energy forming an electromagnetic spectrum. The type of radiation used in the food irradiation process has enough energy to break chemical bonds in molecules or remove tightly bound electrons from atoms, thus creating charged molecules or atoms (ions). This type of radiation is referred to as *ionizing radiation*.

## NATURE OF THE FOOD IRRADIATION INDUSTRY

Food irradiation is one of the current technologies being implemented to preserve and protect the nation's food supply. Irradiating food is not a new idea; it first began in the United States during World War II when the U.S. Army supported experiments with fruits, vegetables, and meats to feed the millions of men and women in uniform (Andress, Delaplane, and Schuler 1998). Scientists were looking for new ways to preserve food without the use of chemicals such as pesticides and herbicides. The irradiation of foods is a technique used to preserve food and rid it of any unwanted pests and pathogenic organisms. The foods are exposed to limited amounts of ionizing radiation. This use of small amounts of radiation can help in improving the quality of the food supply. Food irradiation is said to be a "cold treatment," meaning that it raises the temperature of the food only slightly, minimizing nutrient losses and structural changes in the food.

The irradiation of food can aid in the elimination of disease-causing pathogens, decreasing food-borne illnesses as well as aiding in the preservation and extension of the shelf life of fruits, vegetables, herbs, and meat products. According to the CDC (2005), irradiation is approved for use in more than 35 countries for more than 50 foods worldwide. Food irradiation is being used to

- control mold growth and insect infestations
- inhibit sprouting in root vegetables such as potatoes, onions, and garlic
- increase the shelf life of fruits and vegetables
- destroy parasites that are sometimes found in meat and other food products
- control the growth of microorganisms in herbs and spices
- reduce bacterial contamination and growth on poultry and other meats
- reduce the need for chemicals and pesticides during crop storage and preservation
- provide sterilized food for NASA's astronauts as well as immune-compromised hospital patients

Although the method of food irradiation is a tried and true way to aid in the safety and security of the nation's food supply, irradiated food is still not commonly sold in the United States. When people hear the word *radiation* or anything to do with radiation, their minds instantly conjure up images of nuclear reactor meltdowns such as Chernobyl and Three Mile Island. Large radiation accidents bring illness, death, and despair for many years after the initial release of nuclear material. Workers at nuclear plants during accidents are at risk and are often exposed to massive doses of radiation. Luckily, in the business of food irradiation, nuclear reactors are not part of the process.

**FOOD IRRADIATION PROCESSES**
Food irradiation facilities do not create radioactive wastes nor do they become radioactive. The common features of all food irradiation facilities are the irradiation room and a system to transport the food into and out of the room. The process of food irradiation may utilize any one of three types of irradiators:

- gamma irradiation
- electron beam irradiation
- X-ray irradiation

Gamma irradiation facilities produce ionizing radiation through the use of cobalt-60 (most common) or cesium-137. These radioactive substances give off high-energy protons, referred to as gamma rays. The high-energy gamma rays are capable of penetrating foods to a depth of several feet. When not in use the source is placed in a water bath, which absorbs the radiation. During the food irradiation process, the source is removed from its water bath, into a chamber made of thick concrete. In approximately five years, the cobalt-60 sources decay to 50 percent. They are then shipped back to the

■

## RADIATION ACCIDENT #1

Decatur, Georgia, 1988. Radiation Sterilizers Inc. (RSI) reported a leak of cesium 137 capsules into the water storage pool. The leak could have been detected earlier had the pool water been continuously circulated and monitored through the demineralizer. Significant contamination of pool water remaining unnoticed, contamination of the facility and workers ensued. Cleanup costs exceeded $30 million. (NRC 1989)

■

original nuclear reactor for recharging and reuse. Cesium-137 sources decay to 50 percent in nearly 31 years; replacement is needed infrequently. When shipped for reprocessing or storage, these substances are placed in specialized hardened steel canisters to prevent easy breakage and leakage (CDC 2005; EPA 2008a).

Electron beam irradiators do not use radioactive sources to produce ionizing radiation. The downside to electron beam technology is that it only has the capability of penetrating food approximately an inch in depth. An electron beam consists of current containing high-energy electrons that are launched out of an electron gun. The electron beam generator can easily be switched on and off. Some shielding is still required to protect workers from the high energy beam, not to the extent of thick concrete barriers as needed with gamma irradiators (CDC 2005; EPA 2008a).

X-ray irradiation is the newest technology for irradiation of foods. The X-ray machine produced to irradiate food is just a more powerful version of the one used in everyday applications at hospitals and dental offices. X-rays are produced by bombarding a current of electrons at a metal plate. High-energy photons or X-rays are given off. X-rays are capable of passing through dense and thick foods. The use of X-rays requires much shielding; however, similar to the electron beam technology, no

■

## RADIATION ACCIDENT #2

Maryland, 1991. A worker suffered critical injuries when exposed to ionizing radiation from an electron beam accelerator. The victim developed sores and blisters on his feet, face, and scalp, and lost fingers on both hands. (Bryce 2001)

■

radioactive sources are involved and the X-ray machine can be turned on and off (EPA 2008a).

## POTENTIAL FOR RADIATION EXPOSURE

According to the International Consultative Group on Food Irradiation (1999) and the EPA (2008a), today there are about 170 industrial gamma irradiation facilities operating worldwide, a number of which process foods in addition to other types of products. Most irradiation facilities are used for sterilizing disposable medical and pharmaceutical supplies, and for processing other nonfood items. Facilities are constructed to standard designs with multiple safeguards to protect worker health and safeguard the community should a natural disaster such as an earthquake or tornado occur. Radiation protection at food irradiation facilities consists of several basic elements:

- facility design
- worker training
- operating procedures
- supervision
- regulatory oversight

Most workplace radiation exposures occur due to noncompliance with proper operating and safety procedures, insufficient training, insufficient regulatory control, inadequate maintenance, equipment malfunction, and in a few cases willful violation (Ortiz, Oresegun, and Wheatley 2000). Many activities include certain risks to human beings and the environment. One of the main hazards at irradiation facilities is associated with accidental exposure to ionizing radiation. Per the U.S. Nuclear Regulatory Commission licensing requirements, facilities using radioactive sources must be designed with multiple fail-safe measures to protect personnel and the environment from accidental radiation exposure. The facility design must include effective safety control systems, to include the appropriate amount of shielding necessary to protect workers as well as the public. Safety rests upon worker compliance with operating procedures and proper training. All radiation plants must be regulated, licensed, and inspected by the appropriate regulating bodies (ICFGI 1999).

## REGULATORY AGENCIES: FOOD IRRADIATION AND FACILITY SAFETY

In 2000, Robert E. Robertson, associate director of food and agriculture issues, reported that many federal agencies have regulatory responsibilities related to food irradiation, including the FDA, USDA, the Nuclear Regulatory Commission (NRC), the Occupational Safety and Health Administration, and the Department of Transportation—with the FDA having primary regulatory responsibility for ensuring the safety of irradiated foods. These agencies regulate a range of issues, including the types of food that can be irradiated, the process by which food can be irradiated, the safe use of radiation, the safety of the workers in irradiation facilities, and the safe transportation of radioactive material.

Under the Federal Food, Drug, and Cosmetic Act, the FDA has overall responsibility for regulating the safety of all foods, except for meat, poultry, and some egg products, which are the USDA's responsibility. The FDA's specific responsibilities for food irradiation derive from the Food Additives Amendment of 1958, which gave the agency responsibility for ensuring the safety of food additives. While most food additives are substances that are added to foods, the act specifically defined the source of radiation to be a food additive. Radiation sources (i.e., cobalt-60, electron beams, and X-ray generators) are considered to be food additives because their use could affect the characteristics of food. The FDA's responsibilities for food irradiation include

- determining the safety of the radiation sources used in food processing
- issuing regulations that prescribe the conditions under which foods can be irradiated and the maximum permitted radiation dose
- inspecting the facilities that irradiate food products

Food additive regulations governing the use of food irradiation are initiated either by a petition submitted to the agency or, less often, by the FDA itself. In both instances, the FDA must determine whether the additive is safe under all conditions of permitted use. Once the FDA issues a final regulation, the additive can be used by anyone who adheres to the specified conditions of use. Since 1963, the FDA has approved the use of irradiation for several foods to

- reduce illness-causing microorganisms
- retard product maturation
- meet quarantine requirements for certain insect pests

The USDA's Food Safety and Inspection Service (FSIS) is also responsible for ensuring the safety of certain irradiated foods.

The NRC, the Department of Transportation, and the Occupational Safety and Health Administration have primary responsibilities for environmental and worker safety issues relating to radioactive materials. The NRC is responsible for ensuring that nuclear materials are used safely within irradiation facilities. Food irradiation facilities must meet the NRC's design, operating, management, training, and other requirements and are inspected yearly for compliance. In some instances, the NRC relinquishes regulatory authority to state governments, which must require at least as much protection as the NRC. The NRC and the Department of Transportation share primary responsibility for regulating the transport of radioactive materials. (The U.S. Postal Service, the Department of Energy, and the states are also involved in regulating the transportation of these materials.) The NRC, FSIS, and the Animal and Plant Health Inspection Service regulate aspects of worker protection in facilities that use nuclear materials, irradiate meat and poultry, and irradiate plant products, respectively.

The Occupational Safety and Health Administration regulates worker safety in food irradiation facilities (29 CFR 1910.96). The agency requires that all radiation facilities

- operate under a worker safety program
- establish procedures to protect workers from accidental exposure to radiation
- prominently display caution signs, labels, and signals
- provide each employee with a personal device to measure radiation absorption such as film badges, pocket chambers, pocket dosimeters, or film rings
- ensure that no individual in a restricted area receives higher levels of radiation than those summarized in table 7.1
- ensure that no employee under the age of 18 receives, in one calendar year, a dose of ionizing radiation in excess of 10 percent of the values shown in table 7.1

**Table 7.1.    Allowable radiation dose per quarter**

| Body part(s) | Dose (rems/quarter) |
| --- | --- |
| Whole body: head and trunk; active blood-forming organs; lens of eyes; or gonads | 1.25 |
| Hands and forearms; feet and ankles | 18.75 |
| Skin of whole body | 7.50 |

Source: 29 CFR 1910.1096, OSHA.

## IONIZING RADIATION PROTECTION AND SAFETY

Ionizing radiation sources used in food irradiation processes, like X-rays and gamma rays, can pose a considerable risk to affected workers if not properly controlled. First and foremost, engineering and administrative controls to limit emission of radiation from the source should be put in place. (Refer to table 7.2 for list of possible controls.) The three cardinal mechanisms workers should become familiar with to decrease their personal dose include

- limiting time of exposure
- increasing their distance from the source
- using the appropriate PPE (personal protective equipment), shielding

**Table 7.2.   Controls for ionizing radiation**

| Types of Controls | Accomplished by |
| --- | --- |
| Limiting radiation emissions at source | Limiting the quantity of ionizing material |
| Limiting exposure time | Limiting the employee's time of exposure |
| Prevent access to locations where radiation sources exist | Written procedures to limit exposures |
| Extending distance from the source | Increased distance tends to dilute airborne particles and gases. |
|  | Radiation level decreases with the square of the distance—inverse square law. |
| Shielding | Reducing radiation levels with shielding made of lead, concrete, steel, or water |
| Barriers | Walls and fences keep unauthorized people out. |
| Warnings | Radiation areas should be clearly marked. |
| Evacuation | In case of a release of radioactive materials, employees should be aware of the facility's emergency preparedness and response plan, to include proper evacuation and containment procedures. |
| Security | Physical monitoring and security procedures can be used. |
| Training | Employees working with and around ionizing radiation must be trained on the hazards associated with radiation. |

Source: 29 CFR 1910.26, OSHA.

All in all, employees should only be exposed to the source for the necessary amount of time, since less time exposed equals less dose. While exposed to ionizing radiation, workers should stay as far away from the source as possible, as greater distance equals less dose. Equally important is the usage of the proper amount and type of shielding. In the food irradiation process, as the radiation travels away from the point source, it spreads out, similar to a beam of light, reducing the intensity of the rays. From a point source, dose rate is reduced by the square of the distance. In other words, a person twice as far away from the radiation source would receive one fourth the dose. This relationship can easily be expressed by the Inverse Square Law:

$$I_1(D_1)^2 = I_2(D_2)^2$$

where:
   $I_1$ = intensity 1 at distance $D_1$
   $I_2$ = intensity 2 at distance $D_2$
   $D_1$ = distance 1 from source
   $D_2$ = distance 2 from source

## Radiation Safety Training

Not enough emphasis can be placed on ensuring that appropriate and adequate training is given to employees working near and with ionizing radiation sources.

Often, when working with hazards on a daily basis, people become complacent and forget about the hazard. That may be easy to do with radiation because it is not a hazard that can be seen or detected with the naked eye; however, that does not make its existence banal, but proves it to be most dangerous. Employees will be reminded daily by placards, labeling, and other forms of signage that they indeed work in a radiation area; however, current training and record keeping of the training are paramount to a successful radiation safety program at any food irradiation facility.

## RADIATION ACCIDENTS AT INDUSTRIAL IRRADIATION FACILITIES

A radiation accident involves a nonroutine overexposure to ionizing radiation, as a result either of dispersal of radioactive material or of being too close to a radioactive source. This could occur, for example, following a major accident at a nuclear facility, in industrial or medical settings because of lack of appropriate occupational or patient safety, following loss or theft of radioactive material, or as a result of a deliberate malicious act. Exposure to ionizing radiation can pose a substantial health risk, with the type and level of risk depending on the duration and amount of exposure.

—World Health Organization, 2008

### DID YOU KNOW?

In 1986, the NRC temporarily suspended the license of a Radiation Technology Inc. (RTI) facility in New Jersey to operate for an aggregate of 77 days. The suspension was based on willful violations of NRC regulations involving deliberate bypassing of certain safety systems designed to protect plant workers from accidental exposure to radiation. In addition, company officials pled guilty to two felony convictions for willfully providing false information to NRC and were fined $100,000. The company officials were found guilty and sentenced to two-year prison terms for violating conditions of the license and willfully providing false information to NRC. (NRC 1997)

During the last 57 years there have been, on average, seven registered accidents annually in all countries of the world. There have been even fewer major accidents at industrial irradiation facilities resulting in injuries, fatalities, or environmental contamination in the United States. Most of the accidents happened because proper control procedures were not followed and safety systems were bypassed. None of the recorded events endangered the health of the public at large or environmental safety (Turai and Veress 2001).

According to the General Accounting Office report (2000), in over four decades of transporting the radioactive isotopes used for irradiation, there has never been an accident resulting in the escape of radioactive materials into the environment. In fact in the United States, the NRC has exempted facilities that use radioisotopes (such as cobalt-60 and cesium-137) from having to prepare an environmental impact statement (which is required for nuclear facilities) because it found that these facilities do "not individually or cumulatively have a significant effect on the human environment."

All radioactive materials required for irradiation facilities are transported in lead-shielded steel casks. The containers meet stringent national and international government standards designed to withstand the most severe accidents, including collisions, punctures, and exposure to fire and water. According to the International Consultative Group on Food Irradiation, this excellent safety record exceeds that of other industries shipping hazardous materials, such as toxic chemicals, crude oil, or gasoline.

For additional information, a list of the major scientific and health-related organizations that consider food irradiation to be safe for the public, workers, and the environment is provided:

**U.S. government agencies**

Food and Drug Administration

Department of Agriculture

Public Health Service

Centers for Disease Control and Prevention

### DID YOU KNOW?

In 1987, a licensee (Isomedix Inc.) deliberately bypassed the radiation monitor interlock systems and substituted an administrative procedure for the engineered safeguard provided by the radiation monitor interlock. The substituted cell entry procedure was implemented without NRC review, approval, and incorporation in the license. The alternate procedures did not constitute an entry control device that functioned automatically to prevent inadvertent entry and did not comply with the requirements of 10 CFR Subsection 20.203(c)(6)(i). In addition, the licensee installed jumper cables to bypass ventilation system interlocks that were designed to automatically protect individuals from noxious gases produced as a result of irradiation. (NRC 1989)

### U.S. scientific and health-related organizations

American Dietetic Association

American Medical Association

American Veterinary Medical Association

Council for Agricultural Science and Technology

Institute of Food Technologists

National Association of State Departments of Agriculture

### International scientific and health-related organizations

Food and Agriculture Organization

International Atomic Energy Agency

World Health Organization

Codex Alimentarius Commission

Scientific Committee of the European Union

## REFERENCES AND RECOMMENDED READING

Andress, E., K. Delaplane, and G. Schuler. 1998. *Food Irradiation.* Athens: University of Georgia, Cooperative Extension Service. www.fcs.uga.edu/pubs/current/FDNS-E-3.html (accessed January 5, 2008).

Bryce, S. 2001. Food Irradiation: The Global Agenda. *Nexus Magazine* 8 (2). www.whale.to/m/irradiation.html (accessed January 5, 2008).

Centers for Disease Control and Prevention (CDC). 2005. Food Irradiation. www.cdc.gov/ncidod/dbmd/diseaseinfo/foodirradiation.htm (accessed January 5, 2008).

Food Safety and Inspection Service (FSIS). 1999. Docket 97-076P: Irradiation of Meat and Meat Products: Review of Risk Analysis Issues. www.fsis.usda.gov/OA/topics/irrad-risk.htm (accessed January 5, 2008).

International Consultative Group on Food Irradiation (ICGFI). 1999. Facts about Food Irradiation: A Series of Fact Sheets from the International Consultative Group on Food Irradiation. www.iaea.org/programmes/nafa/d5/public/foodirradiation.pdf (accessed January 5, 2008).

Ortiz, P., M. Oresegun, and J. Wheatley. 2000. Lessons from Major Radiation Accidents. www.irpa.net/irpa10/cdrom/00140.pdf (accessed January 5, 2008).

OSHA 29 CFR 1910.96, Ionizing Radiation Standard.
www.osha.gov/pls/oshaweb/searchresults.category?p_title=&p_text=ionizing+radiation+
(accessed January 5, 2008).

Turai, I., and K. Veress. 2001. Radiation Accidents: Occurrence, Types, Consequences, Medical
Management, and the Lessons to Be Learned. *Central European Journal of Occupational and
Environmental Medicine* 7: 3–14. www.fjokk.hu/cejoem/files/Volume7/Vol7No1/CE01_1-01
.html (accessed January 5, 2008).

United States Environmental Protection Agency (EPA). 2008a. Radiation Protection: Facility
Safety and Environmental Impact. www.epa.gov/radiation/sources/facility_env.html
(accessed January 5, 2008).

———. 2008b. Radiation Protection: Health Effects.
www.epa.gov/radiation/understand/health_effects.html (accessed January 5, 2008).

———. 2008c. Radiation Protection: Radiation and Radioactivity.
www.epa.gov/radiation/understand/index.html (accessed January 5, 2008).

U.S. Food and Drug Administration (FDA). 1998. Image of the Radura.
www.fda.gov/fdac/graphics/1998graphics/iradlogo.gif (accessed January 5, 2008).

U.S. General Accounting Office (GAO). 2000. Food Irradiation: Available Research Indicates
Benefits Outweigh Risks. www.gao.gov/new.items/rc00217.pdf (accessed January 5, 2008).

U.S. Nuclear Regulatory Commission (NRC). 10 CFR 20, Standards for Protection against
Radiation.

———. Office of Nuclear Material Safety and Safeguards. 1989. Information Notice No. 89-82:
Recent Safety Related Incidents at Large Irradiators. www.nrc.gov/reading-rm/doc-
collections/gen-comm/info-notices/1989/in89082.html (accessed January 5, 2008).

———. 1997. NRC SECY-97-019. www.nrc.gov/reading-rm/doc-
collections/commission/secys/1997/secy1997-019/1997-019scy.html (accessed January 5,
2008).

World Health Organization (WHO). 2008. Health Topics: Accidents, Radiation.
www.who.int/topics/accidents_radiation/en/ (accessed January 5, 2008).

# 8

# Ergonomics and Food Manufacturing

**Meatpacking plants have highest rate of repeated-trauma disorders**
Workers in meatpacking plants experienced the highest incidence rate of disorders associated with repeated trauma in 1996. There were 921.6 cases per 10,000 full-time workers in meat packing plants, compared to 33.5 cases per 10,000 workers in private industry as a whole.

—Bureau of Labor Statistics (1999)

In the mid-1980s, the meatpacking and poultry-processing industries began to focus on the problem of work-related musculoskeletal disorders (MSDs). MSDs include injury to the nerves, tendons, muscles, and supporting structures of the hands, wrists, elbows, shoulders, neck, and low back (National Academy of Sciences 2001). In 1986, members of the poultry-processing industry developed a guideline advocating training, the process of ergonomics, and medical intervention as a means to reduce the occurrence of MSDs and their associated costs (DHHS 1997).

In August 1993, OSHA published its *Ergonomics Program Management Guidelines for Meatpacking Plants* (OSHA 1993). The meatpacking guidelines specifically recommended that employers implement an ergonomics process to identify and correct ergonomic-related problems in their work sites. While the meatpacking guidelines were directed primarily to meatpacking plants, many poultry-processing facilities initiated ergonomics programs based upon the recommendations contained in the meatpacking guidelines.

The poultry-processing industry has reduced occupational injuries and illnesses by almost half over the last 10 years (BLS 2002a). Despite these efforts, MSDs are still prevalent in the meatpacking and poultry-processing industry. According to the Bureau of Labor Statistics, of the 3,000 cases (in poultry processing) with days away from work that occurred in 2002, over 30 percent (976 cases) involved MSDs (BLS 2002b, tables 1 and 2). Many meatpacking and poultry-processing jobs involve physically demanding work. Some poultry workers make over 25,000 cuts per day processing chickens and turkeys.

These meatpacking and poultry-processing tasks involve factors, including repetition, force, awkward and static postures, and vibration, that have been identified as increasing the risk of incurring injury. Many of the operations in meatpacking and poultry processing occur with a chilled product or in a cold environment. Cold temperatures in combination with the risk factors may also increase the potential for MSDs to develop (National Academy of Sciences 2001). Excessive exposure to these risk factors can lead to MSDs (DHHS 1997).

In these guidelines, we use the term MSD to refer to a variety of injuries and illnesses that occur from repeated use or overexertion, including

- carpal tunnel syndrome
- tendinitis
- rotator cuff injuries (a shoulder problem)
- epicondylitis (an elbow problem)
- trigger finger
- muscle strains and low back injuries

Employers should consider an MSD to be work-related if an event or exposure in the work environment either caused or contributed to the MSD, or significantly aggravated a preexisting MSD as required by OSHA's recordkeeping rule (29 CFR 1904). For example, when an employee develops carpal tunnel syndrome, the employer needs to looks at the hand activity required for the job and the amount of time spent doing the activity. If an employee develops carpal tunnel syndrome and his or her job requires frequent hand activity, forceful exertions, or sustained awkward hand positions, then the problem may be work-related. If the job requires very little hand activity, then the disorder may not be work-related.

Activities outside of the workplace that involve substantial physical demands may also cause or contribute to MSDs. In addition, development of MSDs may be related to genetic causes, gender, age, and other factors (National Academy of Sciences 2001). Finally, there is evidence that reports of MSDs may be linked to certain psychosocial factors such as job dissatisfaction, monotonous work, and limited job control (National Academy of Sciences 2001; DHHS 1997).

Many changes can be made without significantly increasing costs, and many ergonomic changes result in increased efficiency by reducing the time needed to perform a task. Many poultry-processing companies have already instituted programs that reduce MSDs, reduce workers' compensation costs, and improve efficiency (Jones 1997).

## ERGONOMICS: KEY TERMS

A wide variety of terms are currently used by employers, occupational safety and health professionals, and others in describing ergonomic programs. The following definitions are provided to clarify the terms used by OSHA in the ergonomic program management guidelines.

- **Carpal tunnel syndrome.** Compression of the median nerve as it passes through the carpal tunnel of the wrist, resulting in tingling, pain, or numbness in the thumb and first three fingers.
- **Cumulative trauma.** Injury or illness resulting from repeated and/or excessive demands over time on the musculoskeletal system.
- **Cumulative trauma disorders (CTDs).** The term used for health disorders arising from repeated biomechanical stress due to ergonomic hazards. Other terms that have been used for such disorders include repetitive motion injury, occupational overuse syndrome, and repetitive strain injury.

  CTDs are a class of musculoskeletal disorders involving damage to the tendons, tendon sheaths, synovial lubrication of the tendon sheaths, and the related bones, muscles, and nerves of the hands, wrists, elbows, shoulders, neck, and back. The more frequently occurring occupationally induced disorders in this class include carpal tunnel syndrome, epicondylitis (tennis elbow), tendinitis, tenosynovitis, synovitis, stenosing tenosynovitis of the finger, DeQuervain's disease, and low back pain.
- **Ergonomics.** The study of the relationship between an individual and his or her environmental (such as a workstation). The attempt is to reduce unnecessary stress to the individual through modifications to the environment.
- **Ergonomic hazards.** Workplace conditions that pose a biomechanical stress to the worker. Such hazardous workplace conditions include, but are not limited to, faulty workstation layout, improper work methods, improper tools, excessive tool vibration, and job design problems that include aspects of work flow, line speed, posture and force required, work/rest regimens, and repetition rate. They are also referred to as "stressors."
- **Ergonomic risk factors.** Conditions of a job, process, or operation that contribute to the risk of developing CTDs. Examples include repetitiveness of activity, force required, and awkwardness of posture.
- **Ganglion cyst.** Lesions usually containing thick mucous fluid, normally appearing close to joints and tendon sheaths.
- **Muscle strain.** Overexertion of muscles, resulting in discomfort, pain, or swelling.
- **Repetitive motion.** A single task, motion, or posture that is continuously repeated.
- **Tendinitis.** Inflammation of the tendon and/or tendon sheath at the point of attachment to the muscle, resulting in pain and swelling.

- **Work practices.** A set of procedures for accomplishing a specific task in a manner that reduces or eliminates worker exposure to hazards.

## CUMULATIVE TRAUMA DISORDERS

Experience indicates that workers in the meatpacking, poultry-processing, and other food manufacturing/processing industries with jobs requiring frequent hand exertion may develop cumulative trauma disorders. Cumulative trauma disorders are injuries that develop gradually from related stress to a particular body part. Such disorders are also called "overuse" or "wear-and-tear" repetitive strain disorders. They occur primarily in the upper extremities and include soft tissue injuries such as muscle strain, tendinitis, neuritis, and carpal tunnel syndrome.

A number of things contribute to cumulative trauma disorders. Standing in one position for long periods of time can cause discomfort or strain to muscles of the back and legs, because the muscles remain in a position of contraction without allowing for periods of relaxation of movement. Similarly, the height of a work area may contribute to muscle strain for a very short or very tall individual because he or she may be forced to reach beyond a comfortable point.

Often work requires the active use of the hand or arm, making the upper extremity vulnerable to trauma. The upper extremity includes the shoulder, upper arm, elbow, forearm, wrist, hand, and fingers. The arm and hand move through actions of the joints, muscles, and tendons. Upper extremity movement can range from large, sweeping motion to fine, precise manipulation. Hands and arms work best in a neutral or natural position.

The joints of the upper extremity include the shoulder, elbow, wrist, and fingers. Joints are formed where ligaments connect the end of one bone to another. When joints are twisted beyond their normal range of motion, an injury called a *sprain* occurs.

Muscles are fiber bundles that contract to produce movement. Aching and swelling can result from small strains to muscles. Other injuries may result from the tearing of muscle fibers or from a blow or crush that cause blood to seep out into a large area of the muscle. Such injuries can cause serious damage to the muscle.

Tendons are tough, ropelike structures that attach muscles to bones. When a joint, such as the elbow, is severely stretched or twisted, the tendon fibers can be torn like a rope that is frayed, causing a strain injury or *tendinitis*.

Tendons in the wrist and hand are surrounded by a sheath containing a lubricant called synovial fluid. With overuse this fluid can decrease in amount, causing rubbing or friction between the tendon and the sheath. This condition is called *tenosynovitis*.

Trigger finger is a condition that occurs when the tendon sheath in the finger becomes very swollen and the tendon locks and is unable to move. *Trigger finger* occurs on the palm side of the finger.

The *carpal tunnel* is a very small (2–3 centimeters) tunnel in the wrist. The walls of the tunnel are formed by the bones of the wrist and a tough ligament that wraps around the wrist bones. Tendons to flex the fingers, blood vessels, and a nerve pass through the carpal tunnel from the arm to the hand. If there is swelling of the tendons or other conditions that use up space in the carpal tunnel, the nerve can be pinched or compressed. This can lead to pain, swelling, and numbness in the fingers. As symptoms get worse, weakness and clumsiness will develop in the hand.

Nerves and blood vessels between the neck and shoulder can be compressed, causing numbness and tingling in the hand and fingers. This condition is called *thoracic outlet syndrome.* Thoracic outlet syndrome can frequently be corrected by early diagnosis and an exercise program.

*Ganglion cysts* are smooth, firm, round lumps often noticed on the back of the hand or wrist. These lumps are the most common to form on the hand. Usually they are painless, but a mild aching may be associated with them, especially if tendons are involved. These cysts have fibrous walls filled with mucous-like material. Ganglion cysts sometimes follow injury, but it is not always possible to know the cause of the cysts.

The speed of work may be determined by the speed of a conveyor belt. For example, in chicken processing, the faster the conveyor line, the more frequent is the requirement for the cutting of chicken (the repetition of a specific task). Jobs that require frequent repetition of the task cause muscles to contract frequently, requiring more muscle effort and less recovery time.

Force, for example, required to make a particular cut, either with a knife or scissors, can contribute to cumulative trauma disorders. Increasing the applied force increases muscle effort, decreases circulation to the muscles and causes greater muscle fatigue. Effort required to make a particular cut, either with a knife or scissors, can depend upon the sharpness of the tool. A dull instrument requires more force or exertion and contributes to cumulative trauma disorders.

Continuous muscle contraction can cause tendons in the fingers to swell and become irritated. Forceful gripping may cause pressure on nerves from muscles or tendons, as may repeated movement. Hand and arm motions may include grasping, turning, applying pressure, and pinching. These movements frequently result in stressful hand and wrist positions.

Compression or pressure to nerves (and blood vessels) can also occur when tool handles are squeezed in the palm. Awkward hand motions are sometimes used to separate meat from meat and chicken bones. One hand may hold meat while the other hand is holding the knife to make a specific cut. Scissors can rub on the sides of fingers, causing pressure and compression to nerves of the fingers.

Non-work-related factors may contribute to cumulative trauma disorders. A preexisting condition such as arthritis or a joint injury resulting from sports activity may increase the risk of further injury at work. A worker recovering from illness or a worker with a chronic disease is at increased risk of developing cumulative trauma disorders. Age, sex, and body build can all contribute to cumulative trauma disorders.

## ERGONOMIC SOLUTIONS

Early recognition of problems and treatment of complaints have been very effective in reducing and preventing cumulative trauma disorders. Upon recognition, the ergonomic solutions for meatpacking and poultry processing include engineering changes to workstations and equipment, work practice, personal protective equipment (PPE), and administrative actions. The recommended solutions presented in the following pages (as suggested by OSHA 2004) are not intended to be an exhaustive list, nor do the authors expect or suggest that all of them be used in any given facility. Instead, we encourage members of the food manufacturing industry to develop innovative ergonomic solutions that are appropriate to their facilities. OSHA recommends that employers use engineering techniques, where feasible, as the preferred method of dealing with ergonomic problems in meatpacking and poultry-processing facilities. However, OSHA recognizes that a variety of solutions may be needed in any given facility.

OSHA (2004) recommends that employers train employees to use proper work practices. Proper work practices include proper use and maintenance of pneumatic and power tools, good cutting techniques, proper lifting techniques, and good knife care. Using and maintaining effective PPE is also important. For example, good fitting thermal gloves can help with cold conditions while maintaining the ability to grasp items easily.

Many poultry processors [as well as meatpackers and other food manufacturers] have found that administrative solutions can be used to reduce the duration, frequency, and degree of exposure to risk factors. Some examples of administrative solutions used effectively by [food manufacturing industries] follow:

- Job rotation may alleviate physical fatigue and stress of a particular set of muscles and tendons. To set up a job rotation system, employers typically classify the nature and extent of exertions of each task, and then create a schedule that rotates between high and low repetitions within the line and/or between bending and stretching movements in the same work area or whole plant as appropriate to reduce exposure. Also consider the body parts used and rotate so that body parts used repetitively or in awkward postures can either rest completely or work at slower rates and in better postures. Use a rotation schedule to address tasks considered to be high risk (e.g., using vibrating hand tools or deboning activities) or to minimize exposure to cold.

- Staffing "floaters" provide periodic breaks between scheduled breaks.
- New employees, reassigned employees, employees returning from an extended time off for vacation or some other purpose in [meatpacking or] poultry processing facilities often will need a conditioning, or break-in period to get them accustomed to an activity and strengthen them for the physically demanding work they will be performing. To accommodate this, OSHA recommends that new and reassigned employees be gradually integrated into a full workload. OSHA also recommends that employees be assigned to an experienced trainer for job training and evaluation during the conditioning period.
- Allowing pauses relieves fatigued muscles and allows employees to rest affected muscle groups during that time period.
- Cross-train employees so that sufficient support is available for peak production, to cover breaks, to institute job enlargement programs and to provide additional rotation alternatives.
- Performing routine and preventive maintenance on equipment assures that the equipment is working properly.

When combined with exposure to other risk factors, cold can increase the risk of developing an MSD. Employers typically limit cold exposure by providing a warm, dry area and allowing frequent, short breaks to allow workers to warm up. It is also important to use appropriate clothing and personal protective equipment when working in cold environments (OSHA 2004).

**Examples of Ergonomic Solutions**

The solutions presented below are based on OSHA's Guidelines for Poultry Processing. They are not intended to be an exhaustive list, and are only examples of

## ERGONOMIC NECESSITIES

Work methods should be examined and corrections made where practical and appropriate. Workplaces may need to be redesigned to better accommodate the area to the worker (ergonomic necessities).

ergonomic solutions. Individual food manufacturing operations may decide to use these ideas as a starting point as they look for other innovative methods that will meet their facility's needs.

**Problem 1.** Excessive leaning or reaching is required to access material at a workstation. *Solution:* Remove a section of work surface to allow the employee to get closer to items located at the workstation.

**Problem 2.** Packaging finished products. *Solution:* Employ auto baggers and other mechanisms to place whole food product into packaging, and packages into shipping containers.

**Problem 3.** Unloading the contents of a container. *Solution:* Use a mechanical device that tilts or inverts a container in order to release its contents.

**Problem 4.** The need to drop food parts from a conveyor line operation into separate container. *Solution:* Attach tunnel-type (chute) mechanism to a hole in the workstation surface into which food parts or other items can be dropped and transported.

**Problem 5.** Excessive leaning or reaching is required to access material on a conveyor or slide. *Solution:* Install a mechanical device that directs material on a conveyor or slide.

**Problem 6.** Operations require materials to be weighed. *Solution:* Embed scales to incorporate weighing into the production process to eliminate unnecessary handling of food parts.

**Problem 7.** Manual handling associated with hanging food parts. *Solution:* Use mechanical devices to position and stabilize food parts for processing and transporting to other work areas.

**Problem 8.** In cutting and deboning operations excessive use of force or awkward postures must be used. *Solution:* Install mechanical devices to position and stabilize food parts for processing.

**Problem 9.** Long-term standing or sitting at fixed workstations. *Solution:* Select the most appropriate support device to promote neutral body posture and reduce fatigue during seated, sit/stand, and standing tasks.

**Problem 10.** Excessive forward trunk bending and lifting of the arms when working with food parts. *Solution:* Properly adjust work surfaces.

**Problem 11.** Shelf systems configured in such a manner so as to require excessive lifting, carrying, and awkward postures associated with storage of any item used or produced at a workstation. *Solution:* Racks and shelves designed to optimize manual access to minimize excessive lifting, carrying, and awkward postures associated with storage of any item used or produced at a workstation.

**Problem 12.** Lifting containers causes strain on hands, arms, and back. *Solution:* Tabletop or work surface embedded with roller or ball bearings to reduce friction and force when sliding items.

## REFERENCES AND RECOMMENDED READING

Bureau of Labor Statistics (BLS). 1999. Bulletin 2512: Meat Packing Plants Have the Highest Rate of Repeated-Trauma Disorders. Washington, DC: U.S. Department of Labor.

———. 2002a. Poultry Slaughtering and Processing. Washington, DC: Bureau of Labor Statistics.

———. 2002b. OSHA Docket GE2003-2 Exhibit 4-10: Table 1. Number of median days, and incidence rate of nonfatal occupational injuries and illnesses with days away from work involving musculoskeletal disorders by selected industries, 2002; Table 2. Numbers of nonfatal occupational injuries and illnesses by industry and case types, 2002. Washington, DC: U.S. Department of Labor.

Department of Health and Human Services (DHHS). 1997. *Musculoskeletal Disorders and Workplace Factors: A Critical Review of Epidemiological Evidence for Work-Related Musculoskeletal Disorders of the Neck, Upper Extremity, and Low Back.* Cincinnati, OH: U.S. Department of Health and Human Services, Public Health Service, Centers for Disease Control and Prevention.

Jones, R. J. 1997. Corporate Ergonomics Program of a Large Poultry Processor. *AIHA Journal* 58.

National Academy of Sciences. 2001. *Musculoskeletal Disorders and the Workplace—Low Back and Upper Extremities.* Washington, DC: National Academies Press.

Occupational Safety and Health Administration (OSHA). 1986. *Poultry Industry Task Force. The Medical Ergonomics Training Program: A Guide for the Poultry Industry.* Washington, DC: U.S. Department of Labor.

———. 1993. *Ergonomics Program Management Guidelines for Meatpacking Plants.* Washington, DC: U.S. Department of Labor.

———. 2004. Guidelines for Poultry Processing: Ergonomics for the Prevention of Musculoskeletal Disorders. http://www.osha.gov/ergonomics/guidelines/poultryprocessing.html (accessed August 24, 2007).

# Food Manufacturing and Spanish-Speaking Workers

He who speaks and writes in two languages, has a wide knowledge of his own.
—Johann Wolfgang von Goethe

This chapter deals with employers who are hiring the growing number of Hispanic workers in the food manufacturing industry and demanding bilingual job safety and health training. This chapter will help provide employers with information on how to implement an effective safety and health program for their Hispanic workers.

The number of Hispanics living in the United States has increased approximately 60 percent in the last dozen years, and, at the present time, Hispanics now account for 15 percent of the population, making them the country's largest minority. This growth has occurred not only in border states and Eastern cities, which are typically targets for Hispanic and Latino immigrants, but also in the Midwest and Southeast, where Spanish-speaking arrivals have gravitated toward the food manufacturing industry.

Is the food manufacturing industry taking steps to accommodate the influx of Spanish and other non-English-speaking workers? Hispanics now comprise a significant portion of food manufacturing employees. The Hispanic proportion of the food manufacturing industry workforce has grown rapidly, and this will continue. This raises some of the most important safety and health issues facing the food manufacturing industry.

## SUCCESS WITH HISPANIC OUTREACH

The following is an account submitted to OSHA by Wenner Bread Products Company, an employer that has had success in reaching out to its Spanish-speaking workers.

| | |
|---|---|
| **State:** | New York |
| **Company:** | Wenner Bread Products, Bayport, New York |
| **Industry:** | Bakery Products (SIC Code 2051/NAICS Code 311812) |

Employees:      500
Success Brief:  Wenner Bread Products implemented a Safety Management System
                that addressed language and cultural barriers in its workplace.

### The Problem

Language and cultural barriers involving the company's largely Spanish-speaking workforce contributed to on-the-job accidents and injuries. One of the company's greatest difficulties, before applying for OSHA Voluntary Protection Program (VPP) evaluation help, was that all training and information was shared solely through translation/interpreting.

### The Solution

Upon applying to OSHA for a VPP evaluation, the company's management and safety team worked together to introduce employees to a newly created Safety Management System and the principles of VPP. The company first evaluated its employees' educational levels, job duties, and common injuries, as well as culture and background, and then revised its safety programs and communications materials accordingly.

To better communicate with the numerous Spanish-speaking applicants and employees, the company now conducts separate Spanish-language job interviews and safety orientations for those applicants and new hires with little or no English skills, and makes all educational, operational, and regulatory information available in both Spanish and English. The company has also made its written materials simpler to understand in order to accommodate the different educational levels, and it makes sure that Spanish materials are in the appropriate dialects. The company's daily safety briefings, weekly safety meetings, and weekly safety tips are conducted in both Spanish and English, and medical insurance information booklets are also available in Spanish. Interpreters are provided as needed for employees who require medical appointments to treat work-related injuries.

The company has reduced the cost of this program by utilizing bilingual employees both to translate handouts and manuals and to serve as interpreters for medical appointments and at workplace presentations, including weekly and monthly safety meetings. These employees are then rewarded with points that lead to prizes. Also, in some cases, employees' bilingual skills and services have significantly contributed to their achieving promotions.

### Impact

Since proactively addressing language and cultural barriers in its workplace, the company has experienced a substantial decrease in injuries and illnesses. As

DID YOU KNOW?

Despite their minority status, Hispanics figure disproportionately as the victims of on-the-job injuries and deaths, these dangers being greatest to those who speak and understand English the least. Improving workplace safety thus hinges in part on the ability of occupational safety and health professionals to communicate their vital messages in a foreign language. (Boraiko 2007)

a result, in 2003, the company's overall injury and illness incident rate was 6.6 (compared to 12.7 in General Industry). The company has also experienced improved employee relations with its Spanish-speaking employees since the changes were implemented, as well as a significant increase in productivity and product quality.

The following document is an OSHA Fact Sheet (OSHA 2007a). Please see appendix A for a Spanish version of these guidelines.

## OSHA PROGRAMS TO HELP HISPANIC WORKERS

The Occupational Safety and Health Administration's (OSHA) mission is to assure the safety and health of workers by setting and enforcing standards; providing training, outreach, and education; establishing partnerships; and encouraging continual improvement in workplace safety and health.

In 2001, OSHA created a Hispanic Task Force, which continues to meet regularly, to identify ways to reduce injuries, illnesses, and deaths among Spanish-speaking employers and workers.

### Communication

OSHA created a Spanish page on its website in 2002 and added a Spanish option to its toll-free helpline at 1-800-321-OSHA. The agency surveyed its staff and found more than 180 federal and state OSHA personnel speak Spanish. OSHA has identified regional Hispanic coordinators to oversee Hispanic outreach and continues to recruit Spanish speakers for all offices.

To enhance communication, OSHA's task force recently created an English-to-Spanish and Spanish-to-English glossary of more than 200 words related to safety and health, which is available on the agency's website at www.osha.gov [and included in this chapter]. More than 15 million listeners have heard OSHA radio public service announcements made available to 650 Spanish radio stations emphasizing the importance of safety on the job.

### Translations

OSHA has a dozen publications and ten fact sheets in Spanish on its web page and will continue adding materials as they are translated. Two interactive software packages—eTools—are available in Spanish for free downloading from OSHA's website. One covers commercial sewing and the other addresses construction.

### Training

OSHA's 70 compliance assistance specialists offer workshops and training seminars tailored to local needs for Hispanic workers from landscaping in Georgia to woodworking in Connecticut to training for day laborers in Texas and Illinois. In addition, OSHA has 20 education centers in 35 locations that may offer some safety and health training in Spanish. In addition, 50 nonprofit OSHA training grant recipients are developing safety and health training materials to reach Spanish-speaking workers. For example:

- Georgia Tech Research Corporation is developing training materials focused on hazards faced by concrete delivery truck drivers.
- International Society of Arboriculture is creating an interactive CD-ROM on safe work practices for tree care.
- Pennsylvania Foundry Association is preparing silicosis prevention training for the foundry industry.
- Texas Engineering Extension Services is developing photo training materials on safety and health hazards in oil and gas field operations.
- University of Massachusetts Lowell Research Foundation is making available a Spanish version of the OSHA 10-hour construction outreach training program.

### Partnering Efforts

In addition to the partnership with Mexico and the Mexican consulates, OSHA has developed 10 national alliances and 31 regional alliances that focus on reaching Hispanic workers and businesses on issues such as ergonomics, motor vehicle safety, amputations, falls, and work zone safety. As part of an alliance, OSHA and the National Association of Home Builders are developing a fall protection fact sheet in Spanish.

Wenner Bread in Bayport, New York, one of OSHA's Voluntary Protection Programs participants, has an overall injury and illness rate about half that of others in its industry. The company believes its success is due to an effective safety and health program that includes making training and all educational, operational, and regulatory information available in both Spanish and English. Daily safety briefings and weekly safety meetings are held in both languages. The company uses bilingual employees to translate materials and interpret workplace presentations. Wenner's efforts have also led to improved employee relations, increased productivity, and higher product quality.

Family safety days draw workers for safety training while providing education and entertainment for other family members. In Dallas, OSHA partnered with the Hispanic Contractors and the Mexican Consulate to host a family safety fair earlier this year. Some employers also paid workers who attended at least 10 of the 12 safety sessions.

On March 27, 2004, 200 construction and landscaping workers came to the Family Health and Safety Fair in Hialeah, Florida, cosponsored by OSHA, DOL's Wage and Hour Division, and four additional groups. The fair featured eight safety classes, taught exclusively in Spanish.

### Enforcement

OSHA is investigating whether there is a link between language and cultural barriers and employee deaths. Preliminary findings indicated that about 25 percent of the fatalities the agency investigates are in some way related to language or cultural barriers. The agency will be collecting and analyzing additional data to find ways to eliminate barriers and reduce deaths on the job.

### ENGLISH TO SPANISH OSHA DICTIONARY

Unlike English, Spanish spelling is very phonetic—that is, the letters consistently correspond to the same sounds. Some of those sounds are difficult for an English-speaking tongue to handle, but fortunately, the resulting English accent won't usually get in the way of understanding.

Spanish vowels are very short and pure compared to English ones. Even in unaccented syllables, they are pronounced clearly—completely unlike in English, where unaccented vowels all sound like the vowel in but. *You need to remember that there is no such sound in Spanish.*

It may feel strange to pronounce every vowel, but Spanish won't work without doing it. Spanish vowels correspond to the vowels in the following English words:

a = "father"
e = "bet"
i = "be"
o = "go"
u = "to"

## General OSHA Terms

| English | Spanish |
| --- | --- |

**A**

| | |
| --- | --- |
| abate | corregir |
| abatement | corrección |
| abatement period | período de corrección |
| accident | accidente |
| accident investigation | investigación de accidente |
| Act | Acta, la Ley |
| administrate | administrar |
| administrative law | ley administrativa |
| administrator | administrador |
| affected employee | trabajador (a) afectado (a) |
| agent of the employer | representante del patrón |
| appeal | apelar |
| approve | aprobar |
| area inspection | inspección de zona |
| Area Office | Oficina de Area, Oficina local de OSHA |
| Assistant Secretary | Secretario Adjunto, Secretario Auxiliar, Sub-Secretario, Secretario Asistente |
| assure | confirmar / verificar |
| authorize | autorizar |

**B**

| | |
| --- | --- |
| break the law, rule | quebrantar la ley, la norma |
| bulletin board | tablón de anuncios |

**C**

| | |
| --- | --- |
| catastrophe | catástrofe |
| cause | causa |
| checklist | lista de comprobación |
| chemical hazard communication | comunicación de riesgos químicos |
| citation | citación |
| civil rights | derechos civiles |
| combined version | infracción combinada |
| complainant | querellante, quejista, demandante, denunciante, reclamante |
| complaint | queja, querella, demanda, denuncia, reclamo |

| | |
|---|---|
| complaint inspection | inspección de una queja |
| compliance | cumplimiento, conformidad |
| compliance assistance | asistencia en cumplimiento, assistencia para conformidad |
| Compliance Assistance Specialist | Especialista de Asistencia de Cumplimiento |
| comply | cumplir |
| consult | asesorar |
| consultant | asesor |
| consultation | consulta, aesoría, consultoría |
| consultative section | sección de asistencia técnica |
| contest | impugnación, apelación |
| coordinate | coordinar |
| correction order | orden de corrección |
| CSHO | oficial de cumplimiento de seguridad y salud, Inspector de OSHA |

**D**

| | |
|---|---|
| danger | peligro |
| dangerous | peligroso |
| demonstrate | demostrar |
| department | departamento |
| Department of Labor | Departamento del Trabajo, Ministerio de Trabajo |
| deposition | deposición |
| develop | desarrollar |
| director | director |
| disciplinary | disciplinario |
| discipline | disciplina |
| discrimination | discrimen, discriminación |
| disease | enfermedad |
| display | exhibir, mostrar |
| division | división |

**E**

| | |
|---|---|
| egregious | flagrante |
| emphasis inspection | inspección de énfasis |
| employee/s | trabajador/es, empleado/s |
| employee exposure record | registro de exposición del empleado |
| employee medical record | expediente médico del empleado |

| | |
|---|---|
| employee representative | representante de los trabajadores / empleados |
| employer | empleador, patrono, patrón, jefe, empresario |
| employer representative | representante del empleador, patrono, patrón, jefe, empresario |
| enforce | imponer |
| enforcement activity | acción de vigilancia |
| enforcement section | sección de vigilancia |
| environmental exposure sampling | muestreo de exposición medioambiental |
| establish | establecer |
| establishment | establecimiento |
| evidence | evidencia, prueba |
| exposure | exposición |

**F**

| | |
|---|---|
| fact sheet | hoja informativa, hoja de información, hoja de sucesos, hoja de acontecimientos, hoja de hechos |
| failure to abate | falta de corrección |
| farm | granja |
| farm operation | operación agrícola |
| farm worker | trabajador agrícola |
| fatality | fatalidad, muerte |
| filed | presentado |
| first aid | primeros auxilios |
| first instance violation | infracción de primera instancia |
| fixed place of employment | planta de trabajo fija |
| follow-up inspection | inspección de verificación |

**G**

| | |
|---|---|
| grant | donación, otorgamiento |
| grouped violation | infracción combinada |

**H**

| | |
|---|---|
| hazard/s | riesgo/s, peligro/s |
| hazard communication | comunicación de riesgos |
| health | salud |
| Health Compliance Officer | Oficial de Vigilancia en Salud |
| health hazard | reisgo contra la salud / peligro a la salud |

| | |
|---|---|
| healthy | saludable |
| hearing | audiencia |
| housing | vivienda |
| hygiene | higiene |

**I**

| | |
|---|---|
| illness/es | enfermedad/es |
| imminent danger | peligro inminente |
| implement | poner en práctica |
| improve | mejorar |
| industrial hygiene | higiene industrial |
| industrial hygienist | higienista industrial |
| injury/ies | lesión/es |
| inspect | inspeccionar |
| inspection | inspección |
| inspector | inspector (a), fiscalizador |
| interim order | orden provisional |
| interview | entrevistar, entrevista |
| investigation | investigación |
| issuance | emisión |
| issue | emitir |

**J**

| | |
|---|---|
| judgment | condena |

**L**

| | |
|---|---|
| labor camp | campamento de trabajadores |
| law | ley |
| lawyer | abogado |
| letter of corrective action | carta de acción correctiva |
| local emphasis programs | programas de énfasis local |
| Log of Work-Related Injuries and Illnesses | Diario de lesiones y enfermedades ocupacionales |
| lost workdays | días laborables perdidos |
| lost workdays cases incident rate | índice de incidencia de días laborables perdidos |

**M**

| | |
|---|---|
| mandatory | obligatorio |
| medical treatment | tratamiento médico |

**N**

| | |
|---|---|
| National Emphasis Programs | Programas de énfasis nacional |
| National Office | Oficina Nacional de OSHA |
| noncompliance | incumplimiento |

**O**

| | |
|---|---|
| under oath | bajo juramento |
| occupational | laboral, en el trabajo |
| Occupational Safety and Health Administration | Administración de Seguridad y Salud Ocupacional, Administración de Seguridad y Salud en el Trabajo, Administración de Seguridad y Salud Laboral |
| Occupational Safety and Health Act of 1970 | Ley de Seguridad y Salud Ocupacional de 1970 |
| Occupational Safety and Health Division | División de Seguridad y Salud en el Trabajo |
| offense | ofensa |
| OSH Act | Acta OSH, Ley de OSHA |
| OSHA Inspector | Inspector de OSHA, Oficial de Cumplimiento de OSHA |
| OSHA Strategic Partnership Program | Programa estratégico para asociación con OSHA |
| OSHA web page | página Web de OSHA |
| other than serious violation | infracción no seria |
| owner | dueño |

**P**

| | |
|---|---|
| partnership | asociación |
| penalize | multar, penalizar |
| penalty | multa, penalidad, sanción |
| periodic inspection | inspección periódica |
| permanent | permanente |
| permissible exposure limits | niveles de exposición permitidos |
| person | persona |
| personal exposure samples | muestreo de exposición personal |
| place of employment/workplace | planta de trabajo |
| poster | póster, cartelón |
| prevent | prevenir |
| priority | prioridad |

| | |
|---|---|
| private property | propiedad privada |
| probability | probabilidad |
| program | programa |
| programmed | programado |
| project | proyecto |
| promulgate | promulgar |
| provide | proporcionar / dar |

## Q

| | |
|---|---|
| question | preguntar |

## R

| | |
|---|---|
| random inspection | inspección al azar |
| reasonable | razonable |
| record | registro de exposición del empleado |
| reduce | reducir |
| refer | acuso, imputo |
| refer to | diríjase |
| referral | referencia, remisión |
| referral inspection | inspección de acuso, imputación |
| Regional Office | Oficina Regional |
| regulation | reglamento |
| repeat violation | infracción repetida, violación repetida |
| require | requerir |
| research | investigar |
| respond | responder |
| responsibility | responsabilidad |
| review | revisar, examinar, analizar |
| rights | derechos |
| routine inspection | inspección rutinaria |
| rule | norma, regal |

## S

| | |
|---|---|
| safe | seguro |
| safety | seguridad |
| Safety Compliance Officer | Oficial de Vigilancia en Seguridad |
| safety hazards | peligros a la seguridad |
| scheduling list | lista sistematizada |
| Secretary of Labor | Secretaria del Trabajo |
| section | sección |

| | |
|---|---|
| select | seleccionar |
| Senior Health Compliance Officer | Oficial de Vigilancia en Salud |
| serious | serio, grave |
| serious physical harm | daño físico serio |
| serious violation | infracción seria |
| services | servicios |
| severity | seriedad, severidad |
| standard industrial classification | clasificación industrial estándar |
| standards | normas |
| state programs | programas estatales, planes estatales, oficinas estatales |
| statute | estatuto, ley |
| substantial failure to comply | falta de cumplimiento sustancial |
| suspended penalty | multa suspendida |

**T**

| | |
|---|---|
| technical assistance section | sección de asistencia técnica |
| temporary | temporal |
| threat | amenaza |
| trainer | entrenador |
| training | entrenamiento, adiestramiento, capacitación, formación, educación, instrucción |

**U**

| | |
|---|---|
| unabated violation | infracción no corregida |
| Union | sindicato, unión |
| unprogrammed inspection | inspección no programada |
| U.S. Department of Labor | Departamento del Trabajo de los EE.UU. |

**V**

| | |
|---|---|
| variance | variante |
| verify | verificar |
| violation | infracción, violación |
| Voluntary Protection Programs | Programas de protección voluntaria |

**W**

| | |
|---|---|
| walkaround | recorrido |
| warrant | orden, mandamiento |
| Willful Violation | infracción intencionada, infracción intencional |

| | |
|---|---|
| witness | testigo |
| worker | trabajador (a) |
| working conditions | condiciones laborales |
| workplace | lugares de trabajo, planta de trabajo |

## REFERENCES AND RECOMMENDED READING

Boraiko, A. A. 2007. Translation 101 for Safety Professionals. www.safetycouncil.com/pdf/ 630Boraiko (accessed August 27, 2007).

Hispanic poultry workers injuries are greater than OSHA reports. 2005. *Inside OSHA* 2 (20), October. www.InsideHealthPolicy.com (accessed August 26, 2007).

OSHA. 2007a. Fact Sheet: OSHA Programs to Help Hispanic Workers. www.dol/gov/opa/ media/press/opa/OPA20041371-osha-e.htm (accessed August 27, 1007).

———. 2007b. OSHA General Terms. www.osha.gov/dscp/compliance_assistance/spanish/osha_genral_terms_ensp_freq (accessed August 30, 2007).

# 10

# OSHA Standards, Self-Inspection, and Training

You can hire the best trainers in the world; they can give Einstein-level knowledge to the trainees; the trainees can receive and actually learn from the training; you can see it; you can feel that the results are absolutely positive—all this is good. Right? Well, not necessarily. One thing is certain: even with all the best training in the world, if the employer or trainer does not document the training, then the training *did not* occur—this is certainly the view OSHA and a court of law will take.

Good training, like effective safety compliance, can be easily summarized: **Document! Document! Document!**

—Frank R. Spellman (ODU Occupational Health Lecture 1)

## OSHA STANDARDS

As you are no doubt aware by now, OSHA issues standards or rules to protect workers against hazards on the job. These standards limit the amount of hazardous chemicals workers can be exposed to, require the use of certain safety practices and equipment, and require employers to monitor hazards and maintain records of workplace injuries and illnesses. Employers can be cited and fined if they do not comply with OSHA standards. It is also possible for an employer to be cited under OSHA's "general duty clause," which requires employers to keep their workplaces free of serious recognized hazards. This clause is generally cited when no specific OSHA standard applies to the hazard.

OSHA issues standards for a wide variety of workplace hazards, including

- toxic substances
- harmful physical agents
- electrical hazards
- fall hazards

IN GENERAL,

OSHA standards require that employers

- maintain conditions or adopt practices reasonably necessary and appropriate to protect workers on the job
- be familiar with and comply with standards applicable to their establishments
- ensure that employees have and use personal protective equipment when required for safety and health

- trenching hazards
- hazardous waste
- infectious diseases
- fire and explosion hazards
- dangerous atmospheres
- machine hazards
- process hazards (i.e., those covered under 29 CFR 1910.119)

In addition, as mentioned, where there are no specific OSHA standards, employers must comply with the OSH Act's general duty clause. Again (emphasized here because it is so important and often ignored), the general duty clause, Section 5(a)(1), requires that each employer "furnish . . . a place of employment which is free from recognized hazards that are causing or are likely to cause death or serious physical harm to employees."

In 1970, when the OSH Act was signed into law by President Nixon, Congress gave OSHA two years to develop its book of standards (the OSHA "bible"). Initially, the primary focus was standards related to General Industry (29 CFR 1910). Later, 29 CFR 1926 (construction), 29 CFR 1915 (shipyards), 29 CFR 1928 (agriculture), and others were developed.

It is interesting to note that when OSHA began the standard-writing process, it did not select a team of expert authors to write the original standards. Instead, many of the standards then (many of which survive today, word for word) were "borrowed" from industry, professional organizations, and/or preexisting federal laws. For example, many present-day fire prevention/protection standards emanated from those authored by the National Fire Protection Association (NFPA). Many other standards originated from the American National Standards Institute (ANSI). For example, ANSI Standard B56.1-1969 is the standard for powered industrial trucks; it covers the safety requirements relating to the elements of design, operation, and maintenance of powered industrial trucks. Standards taken from NFPA and ANSI and other industry-wide standard-developing organizations are discussed and substantially agreed upon through consensus by industry. Thus, the name *consensus standard* is attached to this type of standards.

Another source OSHA has used for its standard-making process has been those prepared by professional experts within specific industries and/or professional societies. For example, standards related to safety practices involved in/with compressed gases originated from the Compressed Gas Association (CGA). An example is CGA's Pamphlet P-1, *Safe Handling of Compressed Gases*. This *propriety standard* covers requirements for safe handling, storage, and use of compressed gas cylinders.

OSHA also uses preexisting *federal laws* to fashion safety standards. Some preexisting federal laws enforced by OSHA include

- Federal Supply Contracts Act (Walsh-Healy)
- Federal Service Contracts Act
- Contract Work Hours and Safety Standard Act (Construction Safety Act)
- National Foundation on the Arts and Humanities Act

Standards issued under these acts are now enforced in all industries where they apply.

### The Current Standards-Setting Process

OSHA can begin standards-setting procedures on its own initiative or in response to petitions from other parties, including

- the Secretary of Health and Human Services (HHS)
- the National Institutes for Occupational Safety and Health (NIOSH)
- state and local governments
- nationally recognized standards-producing organizations and employer or labor representatives

Each spring and fall, the Department of Labor publishes in the *Federal Register* a list of all regulations that have work under way. The Regulatory Agenda provides a schedule for the development of standards and regulations so employers, employees, and other interested parties know when they can be expected.

### Development of Standards

OSHA publishes its intention to propose, amend, or revoke a standard in the *Federal Register*, either as (1) a Request for Information or an Advance Notice of Proposed Rulemaking or announcement of a meeting to solicit information to be used in drafting a proposal; or (2) a Notice of Proposed Rulemaking, which sets out the proposed new rule's requirements and provides a specific time for the public to respond. Interested parties may submit written information and evidence. OSHA also may schedule a public hearing to consider various points of view.

After reviewing public comments, evidence, and testimony, OSHA publishes either the full text of any standard amended or adopted and the date it becomes effective, along with an explanation of the standard and the reasons for implementing it; or a determination that no standard or amendment is necessary.

Other federal agencies, such as NIOSH, can recommend standards to OSHA. The OSH Act established the National Institute for Occupational Safety and Health under the Department of HHS as the research agency for occupational safety and health. NIOSH conducts research on various safety and health problems, provides technical assistance to OSHA, and recommends standards for OSHA's adoption.

### Emergency Temporary Standards

Under certain limited conditions, OSHA can set emergency temporary standards that take effect immediately and remain in effect until suspended by a permanent standard. To take such an action, OSHA must determine that workers are in grave danger due to exposure to substances or agents determined to be toxic or physically harmful or to new hazards, and that an emergency standard is necessary to protect them.

OSHA then publishes the emergency temporary standard in the *Federal Register*, where it also serves as a proposed permanent standard. The usual procedures for adopting a permanent standard apply, except that a final ruling should be made within six months.

### Congressional Jurisdiction over OSHA Standards

OSHA submits all final rules to Congress and the General Accounting Office for review. Congress has the authority to repeal a standard by passing a joint resolution under an expedited procedure established by the Small Business Regulatory Enforcement and Fairness Act, or SBREFA, but has done so only once. For the repeal to take effect, the joint resolution must be signed by the president.

### Employer Recourse

An employer who is unable to comply with new requirements or anyone who disagrees with a new standard can

- Petition a court for judicial review
- Request a permanent, temporary, or experimental variance from a standard or regulation
- Apply for an interim order to continue working under existing conditions while OSHA considers a variance request.

### Petitions to Modify or Withdraw Standards or Requirements

Employers or employees may petition OSHA to modify or revoke standards, just as they may petition the agency to develop standards. OSHA continually reviews its standards to keep pace with developing and changing industrial technology.

### Filing a Petition for Judicial Review

Anyone who may be adversely affected by a final or emergency standard may file a petition for judicial review. The objecting party must file the petition within 60 days of the rule's publication with the U.S. Court of Appeals for the circuit in which the petitioner lives or has his or her primary place of business.

Filing an appeals petition will not delay enforcement of a standard, unless the Court of Appeals specifically orders it. OSHA issues permanent standards only after careful consideration of the arguments that were received from the public in written submission and at hearings.

### Guidelines versus Standards

A *guideline* is a tool to assist employers in recognizing and controlling hazards. It is voluntary and not enforceable under the OSH Act. Failure to implement a guideline is not itself a violation of the OSH Act. Nor is failure to implement a guideline by itself a violation of the OSH Act's general duty clause.

Guidelines are more flexible than standards. They can be developed quickly and can be changed easily as new information becomes available with scientific advances. Guidelines make it easier for employers to adopt innovative programs to suit their workplaces, rather than inflexible, one-size-fits-all solutions to issues that may be unique to an industry or facility.

## OVERVIEW OF GENERAL INDUSTRY STANDARDS

In this section, a brief description of general industry standards is provided. The listing provided below is taken from the OSHA webpage *elaws*—OSHA Hazard Awareness Advisor (OSHA 2008). Many of these standards have some application to ensuring the safety and health of food manufacturing workers.

**1904.2–1904.7 Log of occupational injuries and illness.** Provides rules for the maintenance of a log [OSHA 300 Log] and summary of occupational injuries and illness, including the period covered and access and retention of records.

**1904.2 Posting of notice and availability of the Act.** Provides for the posting of notices, which inform employees of their protections and obligations under the OSH Act. It also provides contact information concerning availability of the Act, regulations and applicable standards.

**1904.8 Reporting of fatality or multiple hospitalization incidents.** Provides rules for the reporting of the death of an employee or the in-patient hospitalization of three or more employees resulting from a work-related incident.

**1904.16 Statistical reporting of occupational injuries and illness.** Obligation to maintain a log of occupational injuries and illness upon being notified in writing by the Bureau of Labor Statistics that the employer has been selected to participate in a statistical survey of occupational injuries and illness.

**1910.22 General requirements for walking-working surfaces.** Defines requirements for physical upkeep of the workplace, including those for housekeeping, aisles and passageways, covers and guardrails, and floor loading protection.

**1910.23 Guarding floor and wall openings and holes.** Identifies the need to protect workers against accidental falls through holes in the floor or roof. It also requires guarding of wall openings or access openings.

**1910.24 Fixed industrial stairs.** Provides specifications for the safe design and construction of fixed general industrial stairs.

**1910.25 Portable wood ladders.** Prescribes rules and establishes minimum requirements for the construction, care, and use of the common types of portable wood ladders.

**1910.26 Portable metal ladders.** Prescribes rules and establishes minimum requirements for the construction, care, and use of the common types of portable metal ladders.

**1910.27 Fixed ladders.** Prescribes rules and establishes minimum requirements for the construction of fixed ladders.

**1910.28 Safety requirements for scaffolding.** Prescribes general requirements for the construction, operation, maintenance, and use of scaffolds used in the maintenance of buildings and structures.

**1910.29 Manually propelled mobile ladder stands and scaffolds.** Prescribes rules and requirements for the design, construction, and use of mobile work platforms and rolling scaffolds.

**1910.30 Other working surfaces.** Prescribes rules and requirements for the design, construction, and use of other working surfaces.

**1910.36 General requirements.** Provides general fundamental requirements essential to providing a safe means of egress from fire and like emergencies.

**1910.37 Means of egress, general.** Provides specifications for the design, construction and maintenance of exits, automatic sprinkler systems, fire alarm signaling systems, fire retardant paints, and exit markings.

**1910.38 Employee emergency plans and fire prevention plans.** Prescribes requirements for the designated actions employers and employees must take to ensure employee safety from fire and other emergencies.

**1910.66 Powered platforms for building maintenance.** Prescribes general requirements for the installation of powered platforms dedicated to interior or exterior building maintenance of a specific structure or group of structures.

**1910.67 Vehicle-mounted elevating and rotating work platforms.** Provides specifications for the design and operation of vehicle-mounted elevating and rotating work platforms.

**1910.68 Manlifts.** Provides specifications for the construction, maintenance, inspection, and operation of manlifts in relation to accident hazards.

**1910.94 Ventilation.** Defines criteria for determining whether or not ventilation equipment and systems are required. Also defines specifications for the design, construction, and use of ventilation equipment and systems.

**1910.95 Occupational noise exposure.** Defines when protection against occupational noise must be provided by employer. Defines the means of measuring occupational noise levels, and employee hearing levels. Defines requirements for employer-provided hearing protection equipment and training.

**1910.97 Nonionizing radiation.** Establishes radiation protection guide for exposure to electromagnetic radiation and the requirements for a warning symbol.

**1910.101 Compressed gases.** Defines requirements for inspection, in-plant handling, storage, and utilization of all compressed gases. Also requires safety relief devices for compressed gas containers. Requirements reference Compressed Gas Association [CGA] Pamphlets.

## DID YOU KNOW?

Acetylene is so flammable that the National Electric Code (NEC) has a special designation (its most stringent) for using electrical equipment around acetylene. No other substance falls into this classification!

**1910.102 Acetylene.** Defines requirements for cylinders, piped systems, and generators and filling cylinders for acetylene. Requirements reference Compressed Gas Association Pamphlets.

**1910.103 Hydrogen.** Defines requirements for the design, construction, and testing of gaseous and liquefied hydrogen systems.

**1910.104 Oxygen.** Defines requirements for the design, construction, and testing of bulk oxygen systems.

**1910.105 Nitrous oxide.** Refers to Compressed Gas Association Pamphlet for requirements for design, installation, maintenance, and operation of piped systems for the plant transfer and distribution of nitrous oxide.

**1910.106 Flammable and combustible liquids.** Defines classes of flammable and combustible liquids and their safe storage and use.

**1910.107 Spray finishing using flammable and combustible materials.** Defines requirements for the equipment, processes, and materials used in performing spray finishing using flammable and combustible materials.

**1910.109 Explosives and blasting agents.** Defines requirements for storage, transportation, and use of explosives and blasting agents. Includes provisions for water gel explosives, ammonium nitrate, and small arms ammunition, propellants, and primers.

**1910.110 Storage and handling of liquefied petroleum gases.** Provides requirements for handling and containing and transporting liquefied petroleum gas.

**1910.111 Storage and handling of anhydrous ammonia.** This standard is intended to apply to the design, construction, location, installation, and operation of anhydrous ammonia systems including refrigerated ammonia storage systems.

**1910.119 Process safety management [PSM] of highly hazardous chemicals.** Contains requirements for preventing or minimizing the consequences of catastrophic releases of toxic, reactive, flammable, or explosive chemicals.

**1910.120 Hazardous waste operations and emergency response.** Describes the required procedures and operations for conducting hazardous waste operations and for creating and implementing emergency response plans.

**1910.122–1910.126 Dipping and coating operations.** Defines requirements for construction, ventilation, use of dipping and coating operations. Defines requirements for conditions and operations in immediate vicinity of dipping and coating operations.

**1910.132 General requirements.** Defines general criteria for using personal protective equipment [PPE], and the requirements for its actual use.

**1910.133 Eye and face protection.** Explains when appropriate eye or face protection should be used and the requirements for their construction and use.

**1910.134 Respiratory protection.** Describes conditions under which respiratory protection is required. Also provides specifications for the use of respirators.

**1910.135 Head protection.** Describes conditions under which head protection is required. Also provides criteria for adequate protective helmets.

**1910.136 Foot protection.** Describes conditions under which foot protection is required. Also provides criteria for adequate protective footwear.

**1910.137 Electrical protective equipment.** Provides requirements for design, care, and use for electrical protective equipment.

**1910.138 Hand protection.** Describes conditions under which hand protection is required. Also provides criteria for selecting hand protection.

**1910.141 Sanitation.** Provides requirements for general housekeeping and maintenance of the premises, the keeping and use of the water supply, toilet facilities, change rooms, clothes drying facilities, consumption of food and beverages on the premises, and food handling.

**1910.142 Temporary labor camps.** Regulates the location and facilities of camps that house temporary workers.

**1910.144 Safety color code for marking physical hazards.** Defines the color-coding system for identifying different types of physical hazards.

**1910.145 Specifications for accident prevention signs and tags.** Provides a classification of signs according to use, and the proper wording of signs.

**1910.146 Permit-required confined spaces.** Contains requirements for practices and procedures to protect employees from the hazards of entry into permit-required confined spaces.

**1910.147 The control of hazardous energy (lockout/tagout [LOTO]).** Requires employers to establish a program and utilize procedures for affixing appropriate lockout or tagout devices to energy isolating devices. This will prevent unexpected energization, start-up, or release of stored energy to prevent injury to employees.

**1910.151 Medical services and first aid.** Provides general requirements for medical services and first aid.

**1910.156 Fire brigades.** Contains requirements for the organization, training, and personal protective equipment of fire brigades whenever they are established by an employer.

**1910.157 Portable fire extinguishers.** Provides requirements for the placement, use, maintenance, and testing of portable fire extinguishers provided for the use of employees.

**1910.158 Standpipe and hose systems.** Provides requirements for the placement, use, maintenance, and testing of standpipe and hose systems.

**1910.159 Automatic sprinkler systems.** Provides requirements for the placement, use, maintenance, and testing of automatic sprinkler systems.

**1910.160 Fixed extinguishing systems, general.** Provides requirements for the use and maintenance of general fixed extinguishing systems.

**1910.161 Fixed extinguishing systems, dry chemical.** Provides requirements for the use and maintenance of fixed extinguishing systems using dry chemical as the extinguishing agent.

**1910.162 Fixed extinguishing systems, gaseous agent.** Provides requirements for the use and maintenance of fixed extinguishing systems using a gas as the extinguishing agent.

**1910.163 Fixed extinguishing systems, water spray and foam.** Provides requirements for the use of fixed extinguishing systems using water or foam solution as the extinguishing agent.

**1910.164 Fire detection systems.** Provides requirements for the placement, use, maintenance, and testing of fire detection systems.

**1910.165 Employee alarm systems.** Provides requirements for the use, maintenance, and testing of employee alarm systems.

**1910.169 Air receivers.** Provides requirements for construction, installation, use, and maintenance of compressed air receivers.

**1910.176 Handling materials—general.** Provides requirements for use of mechanical handling equipment, and for materials storage.

**1910.177 Servicing multi-piece and single piece rim wheels.** Provides requirements for the servicing of multi-piece and single piece rim wheels.

**1910.178 Powered industrial trucks.** Contains requirements relating to training of operators, and the design, maintenance, and use of powered industrial trucks.

**1910.179 Overhead and gantry cranes.** Provides requirements for the design, maintenance, use, and inspection of overhead and gantry cranes.

**1910.180 Crawler locomotive and truck cranes.** Provides requirements for the use, inspection, testing, and maintenance of crawler locomotive and truck cranes.

**1910.181 Derricks.** Provides requirements for the use, inspection, testing, and maintenance of derricks.

**1910.183 Helicopters.** Provides requirements for the preparation, use, equipment, and operation of helicopter cranes.

**1910.184 Slings.** Provides requirements for the safe use, inspection, and maintenance of slings made from alloy steel chain, wire rope, metal mesh, natural or synthetic fiber rope, and synthetic web.

**1910.212 General requirements for all machines.** Provides general requirements for machine guarding to protect employees in the machine area from hazards, and for anchoring fixed machinery.

**1910.213 Woodworking machinery.** Provides requirements for general construction of woodworking machines, and for the machine guards. Also addresses the inspection and maintenance of woodworking machinery.

**1910.215 Abrasive wheel machinery.** Provides requirements for the use of guards and flanges on abrasive wheel machinery, and for the design of such guards and flanges.

**1910.216 Mills and calendars in the rubber and plastics industries.** Provides requirements for the installation, design, placement, and use of mills and calendars in the rubber and plastics industries.

**1910.217 Mechanical power presses.** Provides requirements for the guarding, construction, and operation of mechanical power presses. Also provides requirements for the design, construction, and setting and feeding of dies. There are also requirements for the reporting of injuries to employees.

**1910.218 Forging machines.** Provides requirements for the use, installation, inspection, and maintenance of different types of forging machines.

**1910.219 Mechanical power-transmission apparatus.** Provides requirements for the installation, guarding, maintenance, and use of mechanical power-transmission apparatus in general. There are specific provisions for shafting; pulleys; belt, rope, and chain drives; and gears, sprockets, and chains.

**1910.242 Hand and portable powered tools and equipment, general.** Provides general requirements for the safe condition and cleaning of hand and portable powered tools and equipment.

**1910.243 Guarding of portable powered tools.** Provides requirements for the guarding of portable powered tools. Also defines requirements for the design, use, inspection, and maintenance of explosive actuated fastening tools, and power lawnmowers.

**1910.244 Other portable tools and equipment.** Provides requirements for the operation and maintenance of jacks and abrasive blast cleaning nozzles.

**1910.252 General requirements.** Provides general requirements for fire prevention and protection, protection of personnel, and health protection and ventilation when welding, cutting, or brazing.

**1910.253 Oxygen-fuel gas welding and cutting.** Provides requirements for the selection, use, and operation of oxygen-fueled welding equipment.

**1910.254 Arc welding and cutting.** Provides requirements for the selection, use, application, and installation of arc welding equipment.

**1910.255 Resistance welding.** Provides requirements for the installation, use, design, guarding, and maintenance of resistance welding equipment in general. Additional provisions address spot and seam welding machines (nonportable), portable welding machines, and flash welding equipment.

**1910.261 Pulp, paper, and paper-board mills.** Provides rules and safety requirements for the safe design and operation of pulp, paper, and paper-board mills.

**1910.262 Textiles.** Provides general safety requirements, and requirements for the design, guarding, and use of textile equipment, processes, materials, protective equipment, and workroom ventilation.

**1910.263 Bakery equipment.** Provides requirements governing the design, installation, operation, and maintenance of machinery and equipment used within a bakery.

**1910.264 Laundry machinery and operations.** Provides requirements for the design and operation of equipment used in laundries.

**1910.265 Sawmills.** Provides rules and safety requirements for the safe design and operation of sawmills.

**1910.266 Logging operations.** Establishes safety practices, means, methods, and operations for all types of logging, regardless of the end use of wood.

**1910.268 Telecommunications.** Sets forth safety and health standards that apply to the work conditions, practices, means, methods, operations, installations, and processes performed at telecommunications centers and telecommunications field installations.

**1910.269 Electric power generation, transmission, and distribution.** Provides rules governing the operation and maintenance of electric power generation, control, transformation, transmission, and distribution lines and equipment.

**1910.272 Grain handling facilities.** Provides requirements for the control of grain dust fires and explosions, and certain other safety hazards associated with grain handling facilities.

**1910.303 General requirements.** Provides requirements for the installation and use of electrical equipment. Sets forth safety standards that apply to working conditions around electrical equipment.

**1910.304 Wiring design and protection.** Provides requirements for the safe design and protection of electrical wiring.

**1910.305 Wiring methods, components, and equipment for general use.** Provides requirements for the design and installation of electrical equipment for general use, and for methods of wiring.

**1910.306 Specific purpose equipment and installations.** Provides requirements for the design and installation of electrical specific purpose equipment.

**1910.307 Hazardous (classified) locations.** Provides requirements for electrical equipment and wiring in locations which are classified as hazardous depending on the properties of the flammable vapors, liquids or gases, or combustible dusts or fibers which may be present.

**1910.308 Special systems.** Provides requirements for the design and installation of equipment for special electrical systems, and for their wiring methods.

**1910.332 Training.** Provides requirements for training of employees who face a risk of electrical shock that is not reduced by the requirements of sections 1910.303 through 1910.308.

**1910.333 Selection and use of work practices.** Provides requirements on when and how work practices should be employed to prevent electrical shock or other injuries resulting from electrical contacts when work is performed near or on equipment which may be energized.

**1910.334 Use of equipment.** Provides requirements for the handling, inspection, and use of electrical equipment.

**1910.335 Safeguards for personnel protection.** Provides requirements for the use of protective equipment and alerting techniques for the protection of personnel.

**1910.410 Qualifications of dive teams.** Provides requirements for dive team members and for designating person-in-charge.

**1910.420 Safe practices manual.** Provides requirements for developing and maintaining a safe practices manual.

**1910.421 Pre-dive procedures.** Establishes all information that must be provided and all procedures that must be followed prior to each diving operation.

**1910.422 Procedures during dive.** Provides requirements for the procedures and equipment that must be maintained during each diving operation.

**1910.423 Post-dive procedures.** Provides requirements for the procedures that must be performed after each diving operation.

**1910.424 SCUBA diving.** Establishes limits and procedures for engaging in SCUBA diving by an employer.

**1910.425 Surface-supplied air diving.** Establishes limits and procedures for engaging in surface-supplied air diving by an employer.

**1910.426 Mixed-gas diving.** Establishes limits and procedures for engaging in mixed-gas diving by an employer.

**1910.427 Liveboating.** Establishes limits and procedures for engaging in diving operations involving liveboating by an employer.

**1910.430 Equipment.** Provides requirements for the design, construction, maintenance, and use of diving equipment by an employer.

**1910.440 Recordkeeping requirements.** Provides requirements for the recording of occurrences, and the availability of diving records.

**1910.1000 Air contaminants.** The regulation establishes the permissible exposure limits (PEL) for more than 300 chemicals regulated by OSHA. The PEL is based on an eight-hour period of work. Some substances also have a ceiling limit, or a

STEL [short-term exposure limit]. Employers are required to employ engineering controls and substitution to eliminate or reduce the hazard, before placing workers in respiratory protection. Some chemicals may have a "Skin Notation," which means that dermal protection such as barrier creams and gloves are needed to safely work with the chemical.

**1910.1001 Asbestos.** Provides requirements governing the regulation of occupational exposure to asbestos, including exposure limits, personal protective equipment, engineering controls to reduce exposure, hazard communication, medical surveillance, and recordkeeping.

**1910.1002 Coal tar pitch volatiles.** Defines coal tar pitch volatiles as used in 1910.1000 (Table Z-1).

**1910.1003 13 Carcinogens.** Provides requirements for the following chemicals:

- 4-Nitrobiphenyl, Chemical Abstracts Service Register Number (CAS No.) 92933
- alpha-Naphthylamine, CAS No. 134327
- methyl chloromethyl ether, CAS No. 107302
- 3, 3-Dichlorobenzidine (and its salts) CAS No. 91941
- bis-Chloromethyl ether, CAS No. 542881
- beta-Naphthylamine, CAS No. 91598
- Benzidine, CAS No. 92875
- 4-Aminodiphenyl, CAS No. 92671
- Ethyleneimine, CAS No. 151564
- beta-Propiolactone, CAS No. 57578
- 2-Acetylaminofluorene, CAS No. 53963
- 4-Dimethylaminoazo-benzene, CAS No. 60117
- N-Nitrosodimethylamine, CAS No. 62759

**1910.1017 Vinyl chloride.** Provides requirements for the control of employee exposure to vinyl chloride (chloroethene), Chemical Abstracts Service Registry No. 75014. It applies to the manufacture, reaction, packaging, repackaging, storage, handling, or use of vinyl chloride or polyvinyl chloride, but does not apply to the handling or use of fabricated products made of polyvinyl chloride.

**1910.1018 Inorganic arsenic.** This standard applies to all occupational exposures to inorganic arsenic except that this section does not apply to employee exposures in agriculture or resulting from pesticide application, the treatment of wood with preservatives or the utilization of arsenically preserved wood.

**1910.1020 Exposure and medical records access.** Provides employees, their representatives, and the assistant secretary access to employee exposure and medical records.

**1910.1025 Lead.** Provides requirements for the monitoring and control of employee exposure to metallic lead, inorganic lead compounds, and lead soaps.

**1910.1027 Cadmium.** This standard applies to all occupational exposures to cadmium and cadmium compounds, in all forms, and in all industries covered by the Occupational Safety and Health Act, except the construction-related industries, which are covered under 29 CFR 1926.63.

**1910.1028 Benzene.** This section applies to all occupational exposures to benzene, Chemical Abstracts Service Registry No. 71-43-2. This section does not apply to the storage, transportation, distribution, dispensing, sale, or use of gasoline, motor fuels, or other fuels containing benzene subsequent to its final discharge from bulk wholesale storage facilities.

**1910.1029 Coke oven emissions.** This section applies to the control of employee exposure to coke oven emissions, except that this section shall not apply to working conditions with regard to which other federal agencies exercise statutory authority to prescribe or enforce standards affecting occupational safety and health.

**1910.1030 Bloodborne pathogens.** Provides requirements for the control of employee exposure to human blood or other potentially infectious materials.

**1910.1043 Cotton dust.** This section applies to the control of employee exposure to cotton dust in all workplaces where employees engage in yarn manufacturing, engage in slashing and weaving operations, or work in waste houses for textile operations.

**1910.1044 1, 2-dibromo-3-chloropropane.** This section applies to occupational exposure to 1, 2-dibromo-3-chloropropane (DBCP).

**1910.1045 Acrylonitrile.** This section applies to occupational exposures to acrylonitrile (AN), Chemical Abstracts Service Registry No. 000107131.

**1910.1047 Ethylene oxide.** This section applies to occupational exposures to ethylene oxide (EtO), Chemical Abstracts Service Registry No. 75-21-8.

**1910.1048 Formaldehyde.** This standard applies to all occupational exposures to formaldehyde, i.e., from formaldehyde gas, its solutions, and materials that release formaldehyde.

**1910.1050 Methylenedianiline.** This section applies to occupational exposures to MDA, Chemical Abstracts Service Registry No. 101-77-9.

**1910.1052 Methylene chloride.** This applies to occupational exposures to methylene chloride (MC) or dichloromethane (DCM).

**1910.1096 Ionizing radiation.** This applies to occupational exposure to radiation, which includes alpha rays, beta rays, gamma rays, X-rays, neutrons, high-speed electrons, high-speed protons, and other atomic particles; but such term does not include sound or radio waves, or visible light, or infrared or ultraviolet light. (This standard was recently renumbered from 1910.96.)

**1910.1200 Hazard communication.** Provides requirements ensuring that the hazards of all chemicals produced or imported are evaluated, and that information concerning their hazards is transmitted to employers and employees by various means, including container labeling and other forms of warning, material safety data sheets, and employee training.

**1910.1450 Hazardous chemicals in laboratories.** Provides requirements for the control of occupational exposure to hazardous chemicals in laboratories.

## SELF-INSPECTION

We have found that auditing or self-inspection of the workplace is one of the most important tools in the company/industry safety person's toolbox for ensuring regulatory compliance. More importantly, properly conducted workplace self-audits can result in a dramatic reduction of workplace injuries and/or illnesses. Simply, the importance of ensuring that the workplace is free of biological, chemical, and physical hazards *cannot* be overstated. Self-inspection of the workplace will help to ensure these hazards are identified. Once identified, follow-up corrective action is required to remove or mitigate the identified hazards.

Having performed hundreds of self-audits of various industrial workplaces throughout the southeastern United States, we have never found a workplace that an employer or an employee wanted to be unsafe or unhealthful. On the contrary, the mind-set in these workplaces was best characterized as the following: When an employee arrives at the workplace healthy and in one piece, that employee has the right to expect to leave the workplace in the same condition. We believe that providing workers with a safe and healthy workplace is central to their ability to enjoy health, security, and the opportunity to achieve the American dream. Fortunately for most of us, most employers and regulators feel the same about this important commitment—a commitment to ensure a safe place to work for all.

Ensuring workplace safety is an ongoing enterprise. No one wakes up in the morning and waves a magic wand and all hazards automatically disappear. Instead, along with employer/employee commitment to make the workplace safe, a structured protocol for almost constant, ongoing, everyday, every instant self-inspection of the workplace is required. This constant scrutiny of the workplace should be conducted by employers and employees alike. The organizational culture should be one in which employees are trained and encouraged to identify and report workplace hazards; the responsible person in charge should immediately act on these reports and remove the hazards.

Identifying hazards in the workplace is easier if the organization has a designated, dedicated, and experienced safety person responsible for ensuring the safety and health of all employees and protection of the environment.

**DID YOU KNOW?**

Self-inspection is essential if you are to know where probable hazards exist and whether they are under control. (OSHA 2005)

We are not alone in our opinion on the importance of workplace inspections. OSHA (2005) points out that the most widely accepted way to identify hazards is to conduct safety and health inspections (audits), because the only way to be certain of an actual situation is to look at it directly from time to time.

This section includes OSHA *Small Business Handbook* (2005) checklists designed to assist you in self-inspection fact-finding in your workplace.

These checklists can give you some indication of where to begin taking action to make your business safer and more healthful for all of your employees. [Based on more than 30 years of using checklists, we second this statement.] These checklists are by no means all-inclusive and not all of the checklists will apply to your business. You might want to start by selecting the areas that are most critical to your [food manufacturing process], then expanding your self-inspection checklists over time to fully cover all areas that pertain to your business.

Don't spend time with items that have no application to your business. Make sure that each item is seen by you or your designee and leave nothing to memory or chance. Write down what you see or don't see and what you think you should do about it.

Add information from your completed checklists to injury information, employee information, and process and equipment information to build a foundation to help you determine what problems exist. Then, as you use the OSHA standards in your problem-solving process, it will be easier for you to determine the actions

**REMEMBER!**

A checklist is a tool to help, not a definitive statement of what is mandatory. Use checklists only for guidance.

needed to solve these problems. Once the hazards have been identified, institute the control procedures [necessary].

## SELF-INSPECTION SCOPE

Your self-inspections should cover safety and health issues in the following areas:

- **Processing, receiving, shipping, and storage.** Equipment, job planning, layout, heights, floor loads, projection of materials, material handling and storage methods, training for material handling equipment.
- **Building and grounds conditions.** Floors, walls, ceilings, exits, stairs, walkways, ramps, platforms, driveways, aisles.
- **Housekeeping program.** Waste disposal, tools, objects, materials, leakage and spillage, cleaning methods, schedules, work areas, remote areas, storage areas.
- **Electricity.** Equipment, switches, breakers, fuses, switch-boxes, junctions, special fixtures, circuits, insulation, extensions, tools, motors, grounding, national electric code compliance.
- **Lighting.** Type, intensity, controls, conditions, diffusion, location, glare and shadow control.
- **Heating and ventilation.** Type, effectiveness, temperature, humidity, controls, natural and artificial ventilation and exhausting.
- **Machinery.** Points of operation, flywheels, gears, shafts, pulleys, key ways, belts, couplings, sprockets, chains, frames, controls, lighting for tools and equipment, brakes, exhausting, feeding, oiling, adjusting, maintenance, lockout/tagout, grounding, work space, location, purchasing standards.
- **Personnel.** Training, including hazard identification training; experience; methods of checking machines before use; type of clothing; PPE; use of guards; tool storage; work practices; methods for cleaning, oiling, or adjusting machinery.
- **Hand and power tools.** Purchasing standards, inspection, storage, repair, types, maintenance, grounding, use, and handling.
- **Chemicals.** Storage, handling, transportation, spills, disposals, amounts used, labeling, toxicity or other harmful effects, warning signs, supervision, training, protective clothing and equipment, hazard communication requirements.
- **Fire prevention.** Extinguishers, alarms, sprinklers, smoking rules, exits, personnel assigned, separation of flammable materials and dangerous operations, explosion-proof fixtures in hazardous locations, waste disposal, and training of personnel.
- **Maintenance.** Provide regular and preventive maintenance on all equipment used at the work site, recording all work performed on the machinery and by training personnel on the proper care and servicing of the equipment.

- **PPE.** Type, size, maintenance, repair, age, storage, assignment of responsibility, purchasing methods, standards observed, training in care and use, rules of use, method of assignment.
- **Transportation.** Motor vehicle safety, seat belts, vehicle maintenance, safe-driver programs.
- **First-aid program/supplies.** Medical care facilities locations, posted emergency phone numbers, accessible first-aid kits.
- **Evacuation plan.** Establish and practice procedures for an emergency evacuation, e.g., fire, chemical/biological incidents, bomb threat; include escape procedures and routes, critical plant operations, employee accounting following an evacuation, rescue and medical duties, and ways to report emergencies.

## SELF-INSPECTION CHECKLISTS

These checklists are by no means all-inclusive. You should add to them if needed, or delete items that do not apply to your business; however, carefully consider each item and then make your decision. You should refer to the OSHA standards listed in the previous section for specific guidance that may apply to your work situation.

### Employer Posting

- Is the required OSHA Job Safety and Health Protection Poster displayed in a prominent location where all employees are likely to see it?
- Are emergency telephone numbers posted where they can be readily found in case of emergency?
- Where employees may be exposed to toxic substances or harmful physical agents, has appropriate information concerning employee access to medical and exposure records and Material Safety Data Sheets (MSDSs) been posted or otherwise made readily available to affected employees?
- Are signs concerning exit routes, room capacities, floor loading, biohazards, exposures to X-ray, microwave, or other harmful radiation or substances posted where appropriate?
- Is the Summary of Work-Related Injuries and Illnesses (OSHA Form 300A) posted during the months of February, March, and April?

### Recordkeeping

- Are occupational injuries or illnesses, except minor injuries requiring only first aid, recorded as required on the OSHA 300 log?
- Are employee medical records and records of employee exposure to hazardous substances or harmful physical agents up-to-date and in compliance with current OSHA standards?

- ❑ Are employee training records kept and accessible for review by employees, as required by OSHA standards?
- ❑ Have arrangements been made to retain records for the time period required for each specific type of record? (Some records must be maintained for at least 40 years.)
- ❑ Are operating permits and records up-to-date for items such as elevators, air pressure tanks, liquefied petroleum gas tanks, etc.?

### Safety and Health Program

- ❑ Do you have an active safety and health program in operation that includes general safety and health program elements as well as the management of hazards specific to your work site?
- ❑ Is one person clearly responsible for the safety and health program?
- ❑ Do you have a safety committee or group made up of management and labor representatives that meets regularly and reports in writing on its activities?
- ❑ Do you have a working procedure to handle in-house employee complaints regarding safety and health?
- ❑ Are your employees advised of efforts and accomplishments of the safety and health program made to ensure they will have a workplace that is safe and healthful?
- ❑ Have you considered incentives for employees or workgroups who excel in reducing workplace injury/illnesses?

### Medical Services and First Aid

- ❑ Is there a hospital, clinic, or infirmary for medical care near your workplace or is at least one employee on each shift currently qualified to render first aid?
- ❑ Have all employees who are expected to respond to medical emergencies as part of their job responsibilities received first-aid training; had hepatitis B vaccination made available to them; had appropriate training on procedures to protect them from bloodborne pathogens, including universal precautions; and have available and understand how to use appropriate PPE to protect against exposure to bloodborne diseases?
- ❑ If employees have had an exposure incident involving bloodborne pathogens, was an immediate post-exposure medical evaluation and follow-up provided?
- ❑ Are medical personnel readily available for advice and consultation on matters of employees' health?
- ❑ Are emergency phone numbers posted?
- ❑ Are fully supplied first-aid kits easily accessible to each work area, periodically inspected and replenished as needed?

**DID YOU KNOW?**

Pursuant to an OSHA memorandum of July 1, 1992, employees who render first aid only as a collateral duty do not have to be offered pre-exposure hepatitis B vaccine only if the employer includes and implements the following requirements in his/her exposure control plan: (1) the employer must record all first-aid incidents involving the presence of blood or other potentially infectious materials before the end of the work shift during which the first-aid incident occurred; (2) the employer must comply with post-exposure evaluation, prophylaxis, and follow-up requirements of the Bloodborne Pathogens Standard with respect to "exposure incidents," as defined by the standard; (3) the employer must train designated first-aid providers about the reporting procedure; (4) the employer must offer to initiate the hepatitis B vaccination series within 24 hours to all unvaccinated first-aid providers who have rendered assistance in any situation involving the presence of blood or other potentially infectious materials.

- Have first-aid kits and supplies been approved by a physician, indicating that they are adequate for a particular area or operation?
- Is there an eye-wash station or sink available for quick drenching or flushing of the eyes and body in areas where corrosive liquids or materials are handled?

**Fire Protection**
- Is your local fire department familiar with your facility, its location, and specific hazards?
- If you have a fire alarm system, is it certified as required and tested annually?
- If you have interior standpipes and valves, are they inspected regularly?
- If you have outside private fire hydrants, are they flushed at least once a year and on a routine preventive maintenance schedule?
- Are fire doors and shutters in good operating condition?
- Are fire doors and shutters unobstructed and protected against obstructions, including their counterweights?
- Are fire door and shutter fusible links in place?
- Are automatic sprinkler system water control valves, air and water pressure checked periodically as required?
- Is the maintenance of automatic sprinkler systems assigned to responsible persons or to a sprinkler contractor?

❏ Are sprinkler heads protected by metal guards if exposed to potential physical damage?

❏ Is proper clearance maintained below sprinkler heads?

❏ Are portable fire extinguishers provided in adequate number and type and mounted in readily accessible locations?

❏ Are fire extinguishers recharged regularly with this noted on the inspection tag?

❏ Are employees periodically instructed in the use of fire extinguishers and fire protection procedures?

**Personal Protective Equipment and Clothing**

❏ Has the employer determined whether hazards that require the use of PPE (e.g., head, eye, face, hand, or foot protection) are present or are likely to be present?

❏ If hazards or the likelihood of hazards are found, are employers selecting appropriate and properly fitted PPE suitable for protection from these hazards and ensuring that affected employees use it?

❏ Have both the employer and the employees been trained on PPE procedures, i.e., what PPE is necessary for job tasks, when workers need it, and how to properly wear and adjust it?

❏ Are protective goggles or face shields provided and worn where there is any danger of flying particles or corrosive materials?

❏ Are approved safety glasses required to be worn at all times in areas where there is a risk of eye injuries such as punctures, abrasions, contusions, or burns?

❏ Are employees who wear corrective lenses (glasses or contacts) in workplaces with harmful exposures required to wear only approved safety glasses, protective goggles, or use other medically approved precautionary procedures?

❏ Are protective gloves, aprons, shields, or other means provided and required where employees could be cut or where there is reasonably anticipated exposure to corrosive liquids, chemicals, blood, or other potentially infectious materials? See the OSHA Bloodborne Pathogens Standard, 29 CFR 1910.1030(b), for the definition of "other potentially infectious materials."

❏ Are hard hats required, provided, and worn where danger of falling objects exists?

❏ Are hard hats periodically inspected for damage to the shell and suspension system?

❏ Is appropriate foot protection required where there is the risk of foot injuries from hot, corrosive, or poisonous substances, falling objects, crushing, or penetrating actions?

❏ Are approved respirators provided when need? (See 29 CFR 1910.134 for detailed information on respirators or check OSHA's website.)

❑ Is all PPE maintained in a sanitary condition and ready for use?

❑ Are food or beverages consumed only in areas where there is no exposure to toxic material, blood, or other potentially infectious materials?

❑ Is protection against the effects of occupational noise provided when sound levels exceed those of the OSHA Noise standard?

❑ Are adequate work procedures, PPE, and other equipment provided and used when cleaning up spilled hazardous materials?

❑ Are appropriate procedures in place to dispose of or decontaminate PPE contaminated with, or reasonably anticipated to be contaminated with, blood or other potentially infectious materials?

**General Work Environment**

❑ Are all worksites clean, sanitary, and orderly?

❑ Are work surfaces kept dry and appropriate means taken to assure the surfaces are slip-resistant?

❑ Are all spilled hazardous materials or liquids, including blood and other potentially infectious materials, cleaned up immediately and according to proper procedures?

❑ Is combustible scrap, debris, and waste stored safely and removed from the worksite promptly?

❑ Is all regulated waste, as defined in the OSHA Bloodborne Pathogens standard (29 CFR 1910.1030), discarded according to federal, state, and local regulations?

❑ Are accumulations of combustible dust routinely removed from elevated surfaces including the overhead structure of buildings, etc.?

❑ Is combustible dust cleaned up with a vacuum system to prevent suspension of dust particles in the environment?

❑ Is metallic or conductive dust prevented from entering or accumulating on or around electrical enclosures or equipment?

❑ Are covered metal waste cans used for oily or paint-soaked waste?

❑ Are all oil- and gas-fired devices equipped with flame failure controls to prevent flow of fuel if pilots or main burners are not working?

❑ Are paint spray booths, dip tanks, etc., cleaned regularly?

❑ Are the minimum number of toilets and washing facilities provided and maintained in a clean and sanitary fashion?

❑ Are all work areas adequately illuminated?

❑ Are pits and floor openings covered or otherwise guarded?

❑ Have all confined spaces been evaluated for compliance with 29 CFR 1910.146? (Permit required confined spaces.)

**Walkways**

- Are aisles and passageways kept clear and marked as appropriate?
- Are wet surfaces covered with non-slip materials?
- Are holes in the floor, sidewalk, or other walking surface repaired properly, covered, or otherwise made safe?
- Is there safe clearance for walking in aisles where motorized or mechanical handling equipment is operating?
- Are materials or equipment stored in such a way that sharp projections will not interfere with the walkway?
- Are spilled materials cleaned up immediately?
- Are changes of direction or elevations readily identifiable?
- Are aisles or walkways that pass near moving or operating machinery, welding operations, or similar operations arranged so employees will not be subjected to potential hazards?
- Is adequate headroom provided for the entire length of any aisle or walkway?
- Are standard guardrails provided wherever aisle or walkway surfaces are elevated more than 30 inches (76.20 centimeters) above any adjacent floor or the ground?
- Are bridges provided over conveyors and similar hazards?

**Floor and Wall Openings**

- Are floor openings guarded by a cover, a guardrail, or equivalent on all sides (except at stairways or ladder entrances)?
- Are toeboards installed around the edges of permanent floor openings where persons may pass below the opening?
- Are skylight screens able to withstand a load of at least 200 pounds (90.7 kilograms)?
- Is the glass in windows, doors, glass walls, etc., subject to possible human impact, of sufficient thickness and type for the condition of use?
- Are grates or similar type covers over floor openings such as floor drains designed to allow unimpeded foot traffic or rolling equipment?
- Are unused portions of service pits and pits not in use either covered or protected by guardrails or equivalent?
- Are manhole covers, trench covers and similar covers, and their supports designed to carry a truck rear axle load of at least 20,000 pounds (9,072 kilograms) when located in roadways and subject to vehicle traffic?
- Are floor or wall openings in fire-resistant construction provided with doors or covers compatible with the fire rating of the structure and provided with a self-closing feature when appropriate?

### Stairs and Stairways

- ❑ Do standard stair rails or handrails on all stairways have at least four risers?
- ❑ Are all stairways at least 22 inches (55.88 centimeters) wide?
- ❑ Do stairs have landing platforms not less than 30 inches (76.20 centimeters) in the direction of travel and extend 22 inches (55.88 centimeters) in width at every 12 feet (3.6576 meters) or less of vertical rise?
- ❑ Do stairs angle no more than 50 and no less than 30 degrees?
- ❑ Are stairs of hollow-pan type treads and landings filled to the top edge of the pan with solid material?
- ❑ Are step risers on stairs uniform from top to bottom?
- ❑ Are steps slip-resistant?
- ❑ Are stairway handrails located between 30 inches (76.20 centimeters) and 34 inches (86.36 centimeters) above the leading edge of stair treads?
- ❑ Do stairway handrails have at least 3 inches (7.62 centimeters) of clearance between the handrails and the wall or surface they are mounted on?
- ❑ Where doors or gates open directly on a stairway, is a platform provided so the swing of the door does not reduce the width of the platform to less than 21 inches (53.34 centimeters)?
- ❑ Are stairway handrails capable of withstanding a load of 200 pounds (90.7 kilograms), applied within 2 inches (5.08 centimeters) of the top edge in any downward or outward direction?
- ❑ Where stairs or stairways exit directly into any area where vehicles may be operated, are adequate barriers and warnings provided to prevent employees from stepping into the path of traffic?
- ❑ Do stairway landings have a dimension measured in the direction of travel at least equal to the width of the stairway?
- ❑ Is the vertical distance between stairway landings limited to 12 feet (3.6576 meters) or less?

### Elevated Surfaces

- ❑ Are signs posted, when appropriate, showing the elevated surface load capacity?
- ❑ Are surfaces that are elevated more than 30 inches (76.20 centimeters) provided with standard guardrails?
- ❑ Are all elevated surfaces beneath which people or machinery could be exposed to falling objects provided with standard 4-inch (10.16 centimeter) toeboards?
- ❑ Is a permanent means of access and egress provided to elevated storage and work surfaces?
- ❑ Is required headroom provided where necessary?

- Is material on elevated surfaces piled, stacked, or racked in a manner to prevent it from tipping, falling, collapsing, rolling, or spreading?
- Are dock boards or bridge plates used when transferring materials between docks and trucks or railcars?

### Exiting or Egress—Evacuation

- Are all exits marked with an exit sign and illuminated by a reliable light source?
- Are the directions to exits, when not immediately apparent, marked with visible signs?
- Are doors, passageways, or stairways that are neither exits nor access to exits, but could be mistaken for exits, appropriately marked "NOT AN EXIT," "TO BASEMENT," "STOREROOM," etc.?
- Are exit signs labeled with the word "EXIT" in lettering at least 5 inches (12.70 centimeters) high and the stroke of the lettering at least ½-inch (1.2700 centimeters) wide?
- Are exit doors side-hinged?
- Are all exits kept free of obstructions?
- Are at least two means of egress provided from elevated platforms, pits, or rooms where the absence of a second exit would increase the risk of injury from hot, poisonous, corrosive, suffocating, flammable, or explosive substances?
- Are there sufficient exits to permit prompt escape in case of emergency?
- Are special precautions taken to protect employees during construction and repair operations?
- Is the number of exits from each floor of a building and the number of exits from the building itself appropriate for the building occupancy load?
- Are exit stairways that are required to be separated from other parts of a building enclosed by at least 2-hour fire-resistive construction in buildings more than four stories in height, and not less than 1-hour fire-resistive construction elsewhere?
- Where ramps are used as part of required exiting from a building, is the ramp slope limited to 1 foot (0.3048 meter) vertical and 12 feet (3.6576 meters) horizontal?
- Where exiting will be through frameless glass doors, glass exit doors, storm doors, etc., are the doors fully tempered and meet the safety requirements for human impact?

### Exit Doors

- Are doors that are required to serve as exits designed and constructed so that the path of exit travel is obvious and direct?

❑ Are windows that could be mistaken for exit doors made inaccessible by means of barriers or railings?

❑ Are exit doors able to be opened from the direction of exit travel without the use of a key or any special knowledge or effort when the building is occupied?

❑ Is a revolving, sliding, or overhead door prohibited from serving as a required exit door?

❑ Where panic hardware is installed on a required exit door, will it allow the door to open by applying a force of 15 pounds (6.80 kilograms) or less in the direction of the exit traffic?

❑ Are doors on cold storage rooms provided with an inside release mechanism that will release the latch and open the door even if the door is padlocked or otherwise locked on the outside?

❑ Where exit doors open directly onto any street, alley, or other area where vehicles may be operated, are adequate barriers and warnings provided to prevent employees from stepping into the path of traffic?

❑ Are doors that swing in both directions and are located between rooms where there is frequent traffic provided with viewing panels in each door?

**Portable Ladders**

❑ Are all ladders maintained in good condition, joints between steps and side rails tight, all hardware and fittings securely attached, and movable parts operating freely without binding or undue play?

❑ Are non-slip safety feet provided on each metal or rung ladder, and are ladder rungs and steps free of grease and oil?

❑ Are employers prohibited from placing a ladder in front of doors opening toward the ladder unless the door is blocked open, locked, or guarded?

❑ Are employees prohibited from placing ladders on boxes, barrels, or other unstable bases to obtain additional height?

❑ Are employees required to face the ladder when ascending or descending?

❑ Are employees prohibited from using ladders that are broken, have missing steps, rungs, or cleats, broken side rails, or other faulty equipment?

❑ Are employees instructed not to use the top step of ordinary stepladders as a step?

❑ When portable rung ladders are used to gain access to elevated platforms, roofs, etc., does the ladder always extend at least 3 feet (0.9144 meters) above the elevated surface?

❑ Are employees required to secure the base of a portable rung or cleat type ladder to prevent slipping, or otherwise lash or hold it in place?

❑ Are portable metal ladders legibly marked with signs reading "CAUTION—DO NOT USE AROUND ELECTRICAL EQUIPMENT" or equivalent wording?

❑ Are employees prohibited from using ladders as guys, braces, skids, gin poles, or for other than their intended purposes?

❑ Are employees instructed to only adjust extension ladders while standing at a base (not while standing on the ladder or from a position above the ladder)?

❑ Are metal ladders inspected for damage?

❑ Are the rungs of ladders uniformly spaced at 12 inches (30.48 centimeters) center to center?

**Hand Tools and Equipment**

❑ Are all tools and equipment (both company and employee-owned) used at the workplace in good condition?

❑ Are hand tools, such as chisels, punches, etc., which develop mushroomed heads during use, reconditioned or replaced as necessary?

❑ Are broken or fractured handles on hammers, axes, and similar equipment replaced promptly?

❑ Are worn or bent wrenches replaced?

❑ Are appropriate handles used on files and similar tools?

❑ Are employees aware of hazards caused by faulty or improperly used hand tools?

❑ Are appropriate safety glasses, face shields, etc., used while using hand tools or equipment that might produce flying materials or be subject to breakage?

❑ Are jacks checked periodically to ensure they are in good operating condition?

❑ Are tool handles wedged tightly into the heads of all tools?

❑ Are tool cutting edges kept sharp so the tool will move smoothly without binding or skipping?

❑ Are tools stored in a dry, secure location where they cannot be tampered with?

❑ Is eye and face protection used when driving hardened or tempered studs or nails?

**Portable (Power Operated) Tools and Equipment**

❑ Are grinders, saws, and similar equipment provided with appropriate safety guards?

❑ Are power tools used with proper shields, guards, or attachments, as recommended by the manufacturer?

❑ Are portable circular saws equipped with guards above and below the base shoe?

❑ Are circular saw guards checked to ensure that they are not wedged up, leaving the lower portion of the blade unguarded?

❑ Are rotating or moving parts of equipment guarded to prevent physical contact?

❑ Are all cord-connected, electrically operated tools and equipment effectively grounded or of the approved double insulated type?

❏ Are effective guards in place over belts, pulleys, chains, and sprockets on equipment such as concrete mixers, air compressors, etc.?

❏ Are portable fans provided with full guards or screens having openings ½ inch (1.2700 centimeters) or less?

❏ Is hoisting equipment available and used for lifting heavy objects, and are hoist ratings and characteristics appropriate for the task?

❏ Are ground-fault circuit interrupters provided on all temporary electrical 15 and 20 ampere circuits used during periods of construction?

❏ Are pneumatic and hydraulic hoses on power-operated tools checked regularly for deterioration or damage?

### Abrasive Wheel Equipment Grinders

❏ Is the work rest used and kept adjusted to within 1/8 inch (0.3175 centimeter) of the wheel?

❏ Is the adjustable tongue on the top side of the grinder used and kept adjusted to within ¼ inch (0.6350 centimeters) of the wheel?

❏ Do side guards cover the spindle, nut, and flange and 75 percent of the wheel diameter?

❏ Are bench and pedestal grinders permanently mounted?

❏ Are goggles or face shields always worn when grinding?

❏ Is the maximum revolutions per minute (rpm) rating of each abrasive wheel compatible with the rpm rating of the grinder motor?

❏ Are fixed or permanently mounted grinders connected to their electrical supply system with metallic conduit or other permanent wiring method?

❏ Does each grinder have an individual on and off control switch?

❏ Is each electrically operated grinder effectively grounded?

❏ Are new abrasive wheels visually inspected and ring tested before they are mounted?

❏ Are dust collectors and powered exhausts provided on grinders used in operations that produce large amounts of dust?

❏ Are splash guards mounted on grinders that use coolant to prevent the coolant from reaching employees?

❏ Is cleanliness maintained around grinders?

### Power-Actuated Tools

❏ Are employees who operate power-actuated tools trained in their use and required to carry a valid operator's card?

❏ Is each power-actuated tool stored in its own locked container when not being used?

❏ Is a sign at least 7 inches (17.78 centimeters) by 10 inches (25.40 centimeters) with boldface type reading "POWER-ACTUATED TOOL IN USE" conspicuously posted when the tool is being used?

❏ Are power-actuated tools left unloaded until they are ready to be used?

❏ Are power-actuated tools inspected for obstructions or defects each day before use?

❏ Do power-actuated tool operators have and use appropriate PPE such as hard hats, safety goggles, safety shoes, and ear protectors?

**Machine Guarding**

❏ Is there a training program to instruct employees on safe methods of machine operation?

❏ Is there adequate supervision to ensure that employees are following safe machine operating procedures?

❏ Is there a regular program of safety inspection of machinery and equipment?

❏ Is all machinery and equipment kept clean and properly maintained?

❏ Is sufficient clearance provided around and between machines to allow for safe operations, set up and servicing, material handling, and waste removal?

❏ Is equipment and machinery securely placed and anchored to prevent tipping or other movement that could result in personal injury?

❏ Is there a power shut-off switch within reach of the operator's position at each machine?

❏ Can electric power to each machine be locked out for maintenance, repair, or security?

❏ Are the noncurrent-carrying metal parts of electrically operated machines bonded and grounded?

❏ Are foot-operated switches guarded or arranged to prevent accidental actuation by personnel or falling objects?

❏ Are manually operated valves and switches controlling the operation of equipment and machines clearly identified and readily accessible?

❏ Are all emergency stop buttons colored red?

❏ Are all pulleys and belts within 7 feet (2.1336 meters) of the floor or working level properly guarded?

❏ Are all moving chains and gears properly guarded?

❏ Are splash guards mounted on machines that use coolant to prevent the coolant from reaching employees?

❏ Are methods provided to protect the operator and other employees in the machine area from hazards created at the point of operation, ingoing nip points, rotating parts, flying chips, and sparks?

- ❏ Are machine guards secure and arranged so they do not cause a hazard while in use?
- ❏ If special hand tools are used for placing and removing material, do they protect the operator's hands?
- ❏ Are removing drums, barrels, and containers guarded by an enclosure that is interlocked with the drive mechanism so that revolution cannot occur unless the guard enclosure is in place?
- ❏ Do arbors and mandrels have firm and secure bearings, and are they free from play?
- ❏ Are provisions made to prevent machines from automatically starting when power is restored after a power failure or shutdown?
- ❏ Are machines constructed so as to be free from excessive vibration when the largest size tool is mounted and run at full speed?
- ❏ If machinery is cleaned with compressed air, is air pressure controlled and PPE or other safeguards utilized to protect operators and other workers from eye and body injury?
- ❏ Are fan blades protected with a guard having openings no larger than ½ inch (1.2700 centimeters) when operating within 7 feet (2.1336 meters) of the floor?
- ❏ Are saws used for ripping equipped with anti-kickback devices and spreaders?
- ❏ Are radial arm saws so arranged that the cutting head will gently return to the back of the table when released?

## Lockout/Tagout Procedures

- ❏ Is all machinery or equipment capable of movement required to be de-energized or disengaged and blocked or locked out during cleaning, servicing, adjusting, or setting up operations?
- ❏ If the power disconnect for equipment does not also disconnect the electrical control circuit, are the appropriate electrical enclosures identified and is a means provided to ensure that the control circuit can also be disconnected and locked out?
- ❏ Is the locking out of control circuits instead of locking out main power disconnects prohibited?
- ❏ Are all equipment control valve handles provided with a means for locking out?
- ❏ Does the lockout procedure require that stored energy (mechanical, hydraulic, air, etc.) be released or blocked before equipment is locked out for repairs?
- ❏ Are appropriate employees provided with individually keyed personal safety locks?
- ❏ Are employees required to keep personal control of their key(s) while they have safety locks in use?

❑ Is it required that only the employee exposed to the hazard can place or remove the safety lock?

❑ Is it required that employees check the safety of the lockout by attempting a startup after making sure no one is exposed?

❑ Are employees instructed to always push the control circuit stop button prior to re-energizing the main power switch?

❑ Is there a means provided to identify any or all employees who are working on locked-out equipment by their locks or accompanying tags?

❑ Are a sufficient number of accident prevention signs or tags and safety padlocks provided for any reasonably foreseeable repair emergency?

❑ When machine operations, configuration, or size require an operator to leave the control station and part of the machine could move if accidentally activated, is the part required to be separately locked out or blocked?

❑ If equipment or lines cannot be shut down, locked out, and tagged, is a safe job procedure established and rigidly followed?

**Welding, Cutting, and Brazing**

❑ Are only authorized and trained personnel permitted to use welding, cutting, or brazing equipment?

❑ Does each operator have a copy of and follow the appropriate operating instructions?

❑ Are compressed gas cylinders regularly examined for obvious signs of defects, deep rusting, or leakage?

❑ Is care used in handling and storage of cylinders, safety valves, relief valves, etc., to prevent damage?

❑ Are precautions taken to prevent the mixture of air or oxygen with flammable gases, except at a burner or in a standard torch?

❑ Are only approved apparatuses (torches, regulators, pressure reducing valves, acetylene generators, manifolds) used?

❑ Are cylinders kept away from sources of heat and elevators, stairs, or gangways?

❑ Is it prohibited to use cylinders as rollers or supports?

❑ Are empty cylinders appropriately marked and their valves closed?

❑ Are signs posted reading "DANGER, NO SMOKING, MATCHES, OR OPEN LIGHTS," or the equivalent?

❑ Are cylinders, cylinder valves, couplings, regulators, hoses, and apparatuses kept free of oily or greasy substances?

❑ Is care taken not to drop or strike cylinders?

❑ Are regulators removed and valve-protection caps put in place before moving cylinders, unless they are secured on special trucks?

❑ Do cylinders without fixed wheels have keys, handles, or nonadjustable wrenches on stem valves when in service?

❑ Are liquefied gases stored and shipped valve-end up with valve covers in places?

❑ Are employees trained never to crack a fuel gas cylinder valve near sources of ignition?

❑ Before a regulator is removed, is the valve closed and gas released?

❑ Is red used to identify the acetylene (and other fuel-gas) hose, green for the oxygen hose, and black for inert gas and air hoses?

❑ Are pressure-reducing regulators used only for the gas and pressures for which they are intended?

❑ Is open circuit (no-load) voltage of arc welding and cutting machines as low as possible and not in excess of the recommended limits?

❑ Under wet conditions, are automatic controls for reducing no-load voltage used?

❑ Is grounding of the machine frame and safety ground connections of portable machines checked periodically?

❑ Are electrodes removed from the holders when not in use?

❑ Is it required that electric power to the welder be shut off when no one is in attendance?

❑ Is suitable fire extinguishing equipment available for immediate use?

❑ Is the welder forbidden to coil or loop welding electrode cable around his body?

❑ Are wet machines thoroughly dried and tested before use?

❑ Are work and electrode lead cables frequently inspected for wear and damage, and replaced when needed?

❑ Are cable connectors adequately insulated?

❑ When the object to be welded cannot be moved and fire hazards cannot be removed, are shields used to confine heat, sparks, and slag?

❑ Are fire watchers assigned when welding or cutting is performed in locations where a serious fire might develop?

❑ Are combustible floors kept wet, covered with damp sand, or protected by fire-resistant shields?

❑ Are personnel protected from possible electrical shock when floors are wet?

❑ Are precautions taken to protect combustibles on the other side of metal walls when welding is underway?

❑ Are used drums, barrels, tanks and other containers thoroughly cleaned of substances that could explode, ignite, or produce toxic vapors before hot work begins?

❑ Do eye protection, helmets, hand shields, and goggles meet appropriate standards?

❑ Are employees exposed to the hazards created by welding, cutting, or brazing operations protected with PPE and clothing?

- Is a check made for adequate ventilation in and where welding or cutting is performed?
- When working in confined places, are environmental monitoring tests done and means provided for quick removal of welders in case of an emergency?

**Compressors and Compressed Air**

- Are compressors equipped with pressure relief valves and pressure gauges?
- Are compressor air intakes installed and equipped so as to ensure that only clean, uncontaminated air enters the compressor?
- Are air filters installed on the compressor intake?
- Are compressors operated and lubricated in accordance with the manufacturer's recommendations?
- Are safety devices on compressed air systems checked frequently?
- Before a compressor's pressure system is repaired, is the pressure bled off and the system locked out?
- Are signs posted to warn of the automatic starting feature of the compressors?
- Is the belt drive system totally enclosed to provide protection for the front, back, top, and sides?
- Are employees strictly prohibited from directing compressed air towards a person?
- Are employees prohibited from using highly compressed air for cleaning purposes?
- When compressed air is used to clean clothing, are employees trained to reduce the pressure to less than 10 pounds per square inch (psi)?
- When using compressed air for cleaning, do employees wear protective chip guarding and PPE?
- Are safety chains or other suitable locking devices used at couplings of high-pressure hose lines where a connection failure would create a hazard?
- Before compressed air is used to empty containers of liquid, is the safe working pressure of the container checked?
- When compressed air is used with abrasive blast cleaning equipment, is the operating valve a type that must be held open manually?
- When compressed air is used to inflate auto tires, are a clip-on chuck and an in-line regulator preset to 40 psi required?
- Are employees prohibited from using compressed air to clean up or move combustible dust if such action could cause the dust to be suspended in the air and cause a fire or explosion hazard?

### Compressors/Air Receivers

❏ Is every receiver equipped with a pressure gauge and one or more automatic, spring-loaded safety valves?

❏ Is the total relieving capacity of the safety valve able to prevent pressure in the receiver from exceeding the maximum allowable working pressure of the receiver by more than 10 percent?

❏ Is every air receiver provided with a drain pipe and valve at the lowest point for the removal of accumulated oil and water?

❏ Are compressed air receivers periodically drained of moisture and oil?

❏ Are all safety valves tested at regular intervals to determine whether they are in good operating condition?

❏ Is there a current operating permit?

❏ Is the inlet of air receivers and piping systems kept free of accumulated oil and carbonaceous materials?

### Compressed Gas Cylinders

❏ Are cylinders with a water weight capacity over 30 pounds (13.6 kilograms) equipped with a means to connect a valve protector device, or with a collar or recess to protect the valve?

❏ Are cylinders legibly marked to clearly identify the type of gas?

❏ Are compressed gas cylinders stored in areas protected from external heat sources such as flame impingement, intense radiant heat, electric arcs, or high-temperature lines?

❏ Are cylinders located or stored in areas where they will not be damaged by passing or falling objects or subject to tampering by unauthorized persons?

❏ Are cylinders stored or transported in a manner to prevent them from creating a hazard by tipping, falling, or rolling?

❏ Are cylinders containing liquefied fuel gas stored or transported in a position so that the safety relief device is always in direct contact with the vapor space in the cylinder?

❏ Are valve protectors always placed on cylinders when the cylinders are not in use or connected for use?

❏ Are all valves closed off before a cylinder is removed, when the cylinder is empty, and at the completion of each job?

❏ Are low-pressure fuel gas cylinders checked periodically for corrosion, general distortion, cracks, or any other defect that might indicate a weakness or render them unfit for service?

❏ Does the periodic check of low-pressure fuel gas cylinders include a close inspection of the cylinders' bottoms?

**Hoist and Auxiliary Equipment**
❏ Is each overhead electrical hoist equipped with a limit device to stop the hook at its highest and lowest point of safe travel?
❏ Will each hoist automatically stop and hold any load up to 125 percent of its rated load if its actuating force is removed?
❏ Is the rated load of each hoist legibly marked and visible to the operator?
❏ Are stops provided at the safe limits of travel for trolley hoists?
❏ Are the controls of hoists plainly marked to indicate the direction of travel or motion?
❏ Is each cage-controlled hoist equipped with an effective warning device?
❏ Are close-fitting guards or other suitable devices installed on each hoist to ensure that hoist ropes will be maintained in the sheave grooves?
❏ Are all hoist chains or ropes long enough to handle the full range of movement of the application while maintaining two full wraps around the drum at all times?
❏ Are guards provided for nip points or contact points between hoist ropes and sheaves permanently located within 7 feet (2.1336 meters) of the floor, ground, or working platform?
❏ Are employees prohibited from using chains or rope slings that are kinked or twisted and prohibited from using the hoist rope or chain wrapped around the load as a substitute for a sling?
❏ Is the operator instructed to avoid carrying loads above people?

**Industrial Trucks—Forklifts**
❏ Are employees properly trained in the use of the type of industrial truck they operate?
❏ Are only trained personnel allowed to operate industrial trucks?
❏ Is substantial overhead protective equipment provided on high lift rider equipment?
❏ Are the required lift truck operating rules posted and enforced?
❏ Is directional lighting provided on each industrial truck that operates in an area with less than 2 foot-candles per square foot of general lighting?
❏ Does each industrial truck have a warning horn, whistle, gong, or other device that can be clearly heard above normal noise in the areas where it is operated?
❏ Are the brakes on each industrial truck capable of bringing the vehicle to a complete and safe stop when fully loaded?

❑ Does the parking brake of the industrial truck prevent the vehicle from moving when unattended?

❑ Are industrial trucks that operate where flammable gases, vapors, combustible dust, or ignitable fibers may be present approved for such locations?

❑ Are motorized hand and hand/rider trucks designed so that the brakes are applied and power to the drive motor shuts off when the operator releases his or her grip on the device that controls the truck's travel?

❑ Are industrial trucks with internal combustion engines that are operated in buildings or enclosed areas carefully checked to ensure that such operations do not cause harmful concentrations of dangerous gases or fumes?

❑ Are safe distances maintained from the edges of elevated ramps and platforms?

❑ Are employees prohibited from standing or passing under elevated portions of trucks, whether loaded or empty?

❑ Are unauthorized employees prohibited from riding on trucks?

❑ Are operators prohibited from driving up to anyone standing in front of a fixed object?

❑ Are arms and legs kept inside the running lines of the truck?

❑ Are loads handled only within the rated capacity of the truck?

❑ Are trucks in need of repair removed from service immediately?

**Entering Confined Spaces**

❑ Are confined spaces thoroughly emptied of any corrosive or hazardous substances, such as acids or caustics, before entry?

❑ Are all lines to a confined space that contain inert, toxic, flammable, or corrosive materials valved off and blanked or disconnected and separated before entry?

❑ Are all impellers, agitators, or other moving parts and equipment inside confined spaces locked out if they present a hazard?

❑ Is either natural or mechanical ventilation provided prior to confined space entry?

❑ Are appropriate atmospheric tests performed to check for oxygen deficiency, toxic substances, and explosive concentrations in the confined space before entry?

❑ Is adequate illumination provided for the work to be performed in the confined space?

❑ Is the atmosphere inside the confined space frequently tested or continuously monitored during work?

❑ Is there a trained and equipped standby employee positioned outside the confined space, whose sole responsibility is to watch the work in progress, sound an alarm if necessary, and render assistance?

- ❏ Is the standby employee appropriately trained and equipped to handle an emergency?
- ❏ Are employees prohibited from entering the confined space without lifelines and respiratory equipment if there is any question as to the cause of an emergency?
- ❏ Is approved respiratory equipment required if the atmosphere inside the confined space cannot be made acceptable?
- ❏ Is all portable electrical equipment used inside confined spaces either grounded and insulated or equipped with ground fault protection?
- ❏ Are compressed gas bottles forbidden inside the confined space?
- ❏ Before gas welding or burning is started in a confined space, are hoses checked for leaks, torches lighted only outside the confined area, and the confined area tested for an explosive atmosphere each time before a lighted torch is taken into the confined space?
- ❏ If employees will be using oxygen-consuming equipment such as salamanders, torches, furnaces, etc., in a confined space, is sufficient air provided to assure combustion without reducing the oxygen concentration of the atmosphere below 19.5 percent by volume?
- ❏ Whenever combustion-type equipment is used in a confined space, are provisions made to ensure the exhaust gases are vented outside of the enclosure?
- ❏ Is each confined space checked for decaying vegetation or animal matter which may produce methane?
- ❏ Is the confined space checked for possible industrial waste which could contain toxic properties?
- ❏ If the confined space is below ground and near areas where motor vehicles will be operating, is it possible for vehicle exhaust or carbon monoxide to enter the space?

**Environmental Controls**
- ❏ Are all work areas properly illuminated?
- ❏ Are employees instructed in proper first aid and other emergency procedures?
- ❏ Are hazardous substances, blood and other potentially infectious materials, which may cause harm by inhalation, ingestion, or skin absorption or contact, identified?
- ❏ Are employees aware of the hazards involved with the various chemicals they may be exposed to in their work environment, such as ammonia, chlorine, epoxies, caustics, etc.?
- ❏ Is employee exposure to chemicals in the workplace kept within acceptable levels?
- ❏ Can a less harmful method or product be used?
- ❏ Is the work area ventilation system appropriate for the work performed?

❑ Are spray painting operations performed in spray rooms or booths equipped with an appropriate exhaust system?

❑ Is employee exposure to welding fumes controlled by ventilation, use of respirators, exposure time limits, or other means?

❑ Are welders and other nearby workers provided with flash shields during welding operations?

❑ If forklifts and other vehicles are used in buildings or other enclosed areas, are the carbon monoxide levels kept below maximum acceptable concentration?

❑ Has there been a determination that noise levels in the facilities are within acceptable levels?

❑ Are steps being taken to use engineering controls to reduce excessive noise levels?

❑ Are proper precautions being taken when handling asbestos and other fibrous materials?

❑ Are caution labels and signs used to warn of hazardous substances (e.g., asbestos) and biohazards (e.g., bloodborne pathogens)?

❑ Are wet methods used, when practicable, to prevent the emission of airborne asbestos fibers, silica dust, and similar hazardous materials?

❑ Are engineering controls examined and maintained or replaced on a scheduled basis?

❑ Is vacuuming with appropriate equipment used whenever possible rather than blowing or sweeping dust?

❑ Are grinders, saws, and other machines that produce respirable dusts vented to an industrial collector or central exhaust system?

❑ Are all local exhaust ventilation systems designed to provide sufficient air flow and volume for the application, and are ducts not plugged and belts not slipping?

❑ Is PPE provided, used, and maintained wherever required?

❑ Are there written standard operating procedures for the selection and use of respirators where needed?

❑ Are restrooms and washrooms kept clean and sanitary?

❑ Is all water provided for drinking, washing, and cooking potable?

❑ Are all outlets for water that is not suitable for drinking clearly identified?

❑ Are employees' physical capacities assessed before they are assigned to jobs requiring heavy work?

❑ Are employees instructed in the proper manner for lifting heavy objects?

❑ Where heat is a problem, have all fixed work areas been provided with spot cooling or air conditioning?

❑ Are employees screened before assignment to areas of high heat to determine if their health might make them more susceptible to having an adverse reaction?

❑ Are employees working on streets and roadways who are exposed to the hazards of traffic required to wear bright colored (traffic orange) warning vests?

❑ Are exhaust stacks and air intakes located so that nearby contaminated air will not be recirculated within a building or other enclosed area?

❑ Is equipment producing ultraviolet radiation properly shielded?

❑ Are universal precautions observed where occupational exposure to blood or other potentially infectious materials can occur and in all instances where differentiation of types of body fluids or potentially infectious materials is difficult or impossible?

**Flammable and Combustible Materials**

❑ Are combustible scrap, debris, and waste materials (oily rags, etc.) stored in covered metal receptacles and promptly removed from the worksite?

❑ Is proper storage practiced to minimize the risk of fire, including spontaneous combustion?

❑ Are approved containers and tanks used to store and handle flammable and combustible liquids?

❑ Are all connections on drums and combustible liquid piping, vapor- and liquid-tight?

❑ Are all flammable liquids kept in closed containers when not in use (e.g., parts cleaning tanks, pans, etc.)?

❑ Are bulk drums of flammable liquids grounded and bonded to containers during dispensing?

❑ Do storage rooms for flammable and combustible liquids have explosion-proof lights and mechanical or gravity ventilation?

❑ Is liquefied petroleum gas stored, handled, and used in accordance with safe practices and standards?

❑ Are "NO SMOKING" signs posted on liquefied petroleum gas tanks and in areas where flammable or combustible materials are used or stored?

❑ Are liquefied petroleum storage tanks guarded to prevent damage from vehicles?

❑ Are all solvent wastes and flammable liquids kept in fire-resistant, covered containers until they are removed from the worksite?

❑ Is vacuuming used whenever possible rather than blowing or sweeping combustible dust?

❑ Are firm separators placed between containers of combustibles or flammables that are stacked one upon another to ensure their support and stability?

❑ Are fuel gas cylinders and oxygen cylinders separated by distance and fire-resistant barriers while in storage?

- Are fire extinguishers selected and provided for the types of materials in the areas where they are to be used?
  Class A—Ordinary combustible material fires.
  Class B—Flammable liquid, gas, or grease fires.
  Class C—Energized-electrical equipment fires.
- Are appropriate fire extinguishers mounted within 75 feet (22.86 meters) of outside areas containing flammable liquids and within 10 feet (3.048 meters) of any inside storage area for such materials?
- Are extinguishers free from obstructions or blockage?
- Are all extinguishers serviced, maintained, and tagged at intervals not to exceed one year?
- Are all extinguishers fully charged and in their designated places?
- Where sprinkler systems are permanently installed, are the nozzle heads so directed or arranged that water will not be sprayed into operating electrical switchboards and equipment?
- Are safety cans used for dispensing flammable or combustible liquids at the point of use?
- Are all spills of flammable or combustible liquids cleaned up promptly?
- Are storage tanks adequately vented to prevent the development of excessive vacuum or pressure as a result of filling, emptying, or atmosphere temperature changes?
- Are storage tanks equipped with emergency venting that will relieve excessive internal pressure caused by fire exposure?
- Are rules enforced in areas involving storage and use of hazardous materials?

**Hazardous Chemical Exposure**

- Are employees aware of the potential hazards and trained in safe handling practices for situations involving various chemicals stored or used in the workplace such as acids, bases, caustics, epoxies, phenols, etc.?
- Is employee exposure to chemicals kept within acceptable levels?
- Are eye-wash fountains and safety showers provided in areas where corrosive chemicals are handled?
- Are all containers, such as vats, storage tanks, etc., labeled as to their contents (e.g., "CAUSTICS")?
- Are all employees required to use personal protective clothing and equipment when handling chemicals (gloves, eye protection, respirators, etc.)?
- Are flammable or toxic chemicals kept in closed containers when not in use?
- Are chemical piping systems clearly marked as to their content?

❑ Where corrosive liquids are frequently handled in opened containers or drawn from storage vessels or pipelines, are adequate means readily available for neutralizing or disposing of spills or overflows and performed properly and safely?

❑ Are standard operating procedures established and are they being followed when cleaning up chemical spills?

❑ Are respirators stored in a convenient, clean, and sanitary location, and are they adequate for emergencies?

❑ Are employees prohibited from eating in areas where hazardous chemicals are present?

❑ Is PPE used and maintained whenever necessary?

❑ Are there written standard operating procedures for the selection and use of respirators where needed?

❑ If you have a respirator protection program, are your employees instructed on the correct usage and limitations of the respirators?

❑ Are the respirators National Institute of Occupational Safety and Health (NIOSH)-approved for this particular application?

❑ Are they regularly inspected, cleaned, sanitized, and maintained?

❑ If hazardous substances are used in your processes, do you have a medical or biological monitoring system in operation?

❑ Are you familiar with the threshold limit values or permissible exposure limits of airborne contaminants and physical agents used in your workplace?

❑ Have appropriate control procedures been instituted for hazardous materials, including safe handling practices and the use of respirators and ventilation systems?

❑ Whenever possible, are hazardous substances handled in properly designed and exhausted booths or similar locations?

❑ Do you use general dilution or local exhaust ventilation systems to control dusts, vapors, gases, fumes, smoke, solvents, or mists that may be generated in your workplace?

❑ Is operational ventilation equipment provided for removal of contaminants from production grinding, buffing, spray painting, and/or vapor degreasing?

❑ Do employees complain about dizziness, headaches, nausea, irritation, or other factors of discomfort when they use solvents or other chemicals?

❑ Is there a dermatitis problem? Do employees complain about dryness, irritation, or sensitization of the skin?

❑ Have you considered having an industrial hygienist or environmental health specialist evaluate your operation?

❑ If internal combustion engines are used, is carbon monoxide kept within acceptable levels?

❏ Is vacuuming used rather than blowing or sweeping dust whenever possible for cleanup?

❏ Are materials that give off toxic, asphyxiant, suffocating, or anesthetic fumes stored in remote or isolated locations when not in use?

**Hazardous Substances Communication**

❏ Is there a list of hazardous substances used in your workplace and an MSDS readily available for each hazardous substance used?

❏ Is there a current written exposure control plan for occupational exposure to bloodborne pathogens and other potentially infectious materials, where applicable?

❏ Is there a written hazard communication program dealing with MSDSs, labeling, and employee training?

❏ Is each container for a hazardous substance (i.e., vats, bottles, storage tanks, etc.) labeled with product identity and a hazard warning (communication of the specific health hazards and physical hazards)?

❏ Is there an employee training program for hazardous substances that includes
  ❏ an explanation of what an MSDS is and how to use and obtain one
  ❏ MSDS contents for each hazardous substance or class of substances
  ❏ explanation of "A Right to Know"
  ❏ identification of where an employee can see the written hazard communication program
  ❏ location of physical and health hazards in particular work areas and the specific protective measures to be used
  ❏ details of the hazard communication program, including how to use the labeling system and MSDSs

❏ Does the employee training program on the bloodborne pathogens standard contain the following elements:
  ❏ an accessible copy of the standard and an explanation of its contents
  ❏ a general explanation of the epidemiology and symptoms of bloodborne diseases
  ❏ an explanation of the modes of transmission of bloodborne pathogens
  ❏ an explanation of the employer's exposure control plan and the means by which employees can obtain a copy of the written plan
  ❏ an explanation of the appropriate methods for recognizing tasks and the other activities that may involve exposure to blood and other potentially infectious materials
  ❏ an explanation of the use and limitations of methods that will prevent or reduce exposure, including appropriate engineering controls, work practices, and PPE

❏ information on the types, proper use, location, removal, handling, decontamination, and disposal of PPE

❏ an explanation of the basis for selection of PPE

❏ information on the hepatitis B vaccine

❏ information on the appropriate actions to take and persons to contact in an emergency involving blood or other potentially infectious materials

❏ an explanation of the procedure to follow if an exposure incident occurs, including the methods of reporting the incident and the medical follow-up that will be made available

❏ information on post-exposure evaluations and follow-up

❏ an explanation of signs, labels, and color coding

❏ Are employees trained in

❏ how to recognize tasks that might result in occupational exposure

❏ how to use work practice, engineering controls, and PPE, and their limitations

❏ how to obtain information on the types, selection, proper use, location, removal, handling, decontamination, and disposal of PPE

❏ whom to contact and what to do in an emergency

### Electrical

❏ Do you require compliance with OSHA standards for all contract electrical work?

❏ Are all employees required to report any obvious hazard to life or property in connection with electrical equipment or lines as soon as possible?

❏ Are employees instructed to make preliminary inspections and/or appropriate tests to determine conditions before starting work on electrical equipment or lines?

❏ When electrical equipment or lines are to be serviced, maintained, or adjusted, are necessary switches opened, locked out, or tagged, whenever possible?

❏ Are portable electrical tools and equipment grounded or of the double insulated type?

❏ Are electrical appliances such as vacuum cleaners, polishers, vending machines, etc., grounded?

❏ Do extension cords have a grounding conductor?

❏ Are multiple plug adaptors prohibited?

❏ Are ground-fault circuit interrupters installed on each temporary 15 or 20 ampere, 120 volt alternating current (AC) circuit at locations where construction, demolition, modifications, alterations, or excavations are being performed?

❏ Are all temporary circuits protected by suitable disconnecting switches or plug connectors at the junction with permanent wiring?

- Do you have electrical installations in hazardous dust or vapor areas? If so, do they meet the National Electrical Code (NEC) for hazardous locations?
- Are exposed wiring and cords with frayed or deteriorated insulation repaired or replaced promptly?
- Are flexible cords and cables free of splices or taps?
- Are clamps or other securing means provided on flexible cords or cables at plugs, receptacles, tools, equipment, etc., and is the cord jacket securely held in place?
- Are all cord, cable, and raceway connections intact and secure?
- In wet and damp locations, are electrical tools and equipment appropriate for the use or location or otherwise protected?
- Is the location of electrical power lines and cables (overhead, underground, under floor, other side of walls, etc.) determined before digging, drilling, or similar work is begun?
- Are metal measuring tapes, ropes, hand-lines, or similar devices with metallic thread woven into the fabric prohibited where they could come in contact with energized parts of equipment or circuit conductors?
- Is the use of metal ladders prohibited where the ladder or the person using the ladder could come in contact with energized parts of equipment, fixtures, or circuit conductors?
- Are all disconnecting switches and circuit breakers labeled to indicate their use or equipment served?
- Are disconnecting means always opened before fuses are replaced?
- Do all interior wiring systems include provisions for grounding metal parts of electrical raceways, equipment, and enclosures?
- Are all electrical raceways and enclosures securely fastened in place?
- Are all energized parts of electrical circuits and equipment guarded against accidental contact by approved cabinets or enclosures?
- Are all energized parts of electrical circuits and equipment guarded against accidental contact by approved cabinets or enclosures?
- Is sufficient access and working space provided and maintained around all electrical equipment to permit ready and safe operations and maintenance?
- Are all unused openings (including conduit knockouts) in electrical enclosures and fittings closed with appropriate covers, plugs, or plates?
- Are electrical enclosures such as switches, receptacles, junction boxes, etc., provided with tight-fitting covers or plates?
- Are disconnecting switches for electrical motors in excess of two horsepower able to open the circuit when the motor is stalled without exploding? (Switches must be horsepower rated equal to or in excess of the motor rating.)

❑ Is low voltage protection provided in the control device of motors driving machines or equipment that could cause injury from inadvertent starting?

❑ Is each motor disconnecting switch or circuit breaker located within sight of the motor control device?

❑ Is each motor located within sight of its controller or is the controller disconnecting means able to be locked open or is a separate disconnecting means installed in the circuit within sight of the motor?

❑ Is the controller for each motor that exceeds two horsepower rated equal to or above the rating of the motor it serves?

❑ Are employees who regularly work on or around energized electrical equipment or lines instructed in cardiopulmonary resuscitation (CPR)?

❑ Are employees prohibited from working alone on energized lines or equipment over 600 volts?

**Noise**

❑ Are there areas in the workplace where continuous noise levels exceed 85 decibels?

❑ Is there an ongoing preventive health program to educate employees in safe levels of noise, exposures, effects of noise on their health, and the use of personal protection?

❑ Have work areas where noise levels make voice communication between employees difficult been identified and posted?

❑ Are noise levels measured with a sound level meter or an octave band analyzer and are records being kept?

❑ Have engineering controls been used to reduce excessive noise levels? Where engineering controls are determined to be infeasible, are administrative controls (i.e., worker rotation) being used to minimize individual employee exposure to noise?

❑ Is approved hearing protective equipment (noise attenuating devices) available to every employee working in noisy areas?

❑ Have you tried isolating noisy machinery from the rest of your operation?

❑ If you use ear protectors, are employees properly fitted and instructed in their use?

❑ Are employees in high noise areas given periodic audiometric testing to ensure that you have an effective hearing protection system?

**Identification of Piping Systems**

❑ When nonpotable water is piped through a facility, are outlets or taps posted to alert employees that the water is unsafe and not to be used for drinking, washing, or other personal use?

❑ When hazardous substances are transported through above-ground piping, is each pipeline identified at points where confusion could introduce hazards to employees?

❑ When pipelines are identified by color painted bands or tapes, are the bands or tapes located at reasonable intervals and at each outlet, valve, or connection, and are all visible parts of the line so identified?

❑ When pipelines are identified by color, is the color code posted at all locations where confusion could introduce hazards to employees?

❑ When the contents of pipelines are identified by name or name abbreviation, is the information readily visible on the pipe near each valve or outlet?

❑ When pipelines carrying hazardous substances are identified by tags, are the tags constructed of durable materials, the message printed clearly and permanently, and are tags installed at each valve or outlet?

❑ When pipelines are heated by electricity, steam, or other external source, are suitable warning signs or tags placed at unions, valves, or other serviceable parts of the system?

## Materials Handling

❑ Is there safe clearance for equipment through aisles and doorways?

❑ Are aisleways permanently marked and kept clear to allow unhindered passage?

❑ Are motorized vehicles and mechanized equipment inspected daily or prior to use?

❑ Are vehicles shut off and brakes set prior to loading or unloading?

❑ Are containers of liquid combustibles or flammables, when stacked while being moved, always protected by dunnage (packing material) sufficient to provide stability?

❑ Are dock boards (bridge plates) used when loading or unloading operations are taking place between vehicles and docks?

❑ Are trucks and trailers secured from movement during loading and unloading operations?

❑ Are dock plates and loading ramps constructed and maintained with sufficient strength to support imposed loading?

❑ Are hand trucks maintained in safe operating condition?

❑ Are chutes equipped with sideboards of sufficient height to prevent the materials being handled from falling off?

❑ Are chutes and gravity roller sections firmly placed or secured to prevent displacement?

❑ Are provisions made to brake the movement of the handled materials at the delivery end of rollers or chutes?

- Are pallets usually inspected before being loaded or moved?
- Are safety latches and other devices being used to prevent slippage of materials off of hoisting hooks?
- Are securing chains, ropes, chockers, or slings adequate for the job?
- Are provisions made to ensure that no one is below when hoisting material or equipment?
- Are MSDSs available to employees handling hazardous substances?

**Control of Harmful Substances by Ventilation**
- Is the volume and velocity of air in each exhaust system sufficient to gather the dusts, fumes, mists, vapors, or gases to be controlled, and to convey them to a suitable point of disposal?
- Are exhaust inlets, ducts, and plenums designed, constructed, and supported to prevent collapse or failure of any part of the system?
- Are clean-out ports or doors provided at intervals not to exceed 12 feet (3.6576 meters) in all horizontal runs of exhaust ducts?
- Where two or more different operations are being controlled through the same exhaust system, could the combination of substances involved create a fire, explosion, or chemical reaction hazard in the duct?
- Is adequate makeup air provided to areas where exhaust systems are operating?
- Is the source point for makeup air located so that only clean, fresh air, free of contaminants, will enter the work environment?
- Where two or more ventilation systems serve a work area, is their operation such that one will not offset the functions of the other?

**Sanitizing Equipment and Clothing**
- Is required personal protective clothing or equipment able to be cleaned and disinfected easily?
- Are employees prohibited from interchanging personal protective clothing or equipment, unless it has been properly cleaned?
- Are machines and equipment that process, handle, or apply materials that could injure employees cleaned and/or decontaminated before being overhauled or placed in storage?
- Are employees prohibited from smoking or eating in any area where contaminants are present that could be injurious if ingested?
- When employees are required to change from street clothing into protective clothing, is a clean change room with a separate storage facility for street and protective clothing provided?

❑ Are employees required to shower and wash their hair as soon as possible after a known contact with a carcinogen has occurred?

❑ When equipment, materials, or other items are taken into or removed from a carcinogen-regulated area, is it done in a manner that will not contaminate non-regulated areas or the external environment?

## SAFETY AND HEALTH TRAINING

Throughout this text, one of the major points we make and re-make has been to emphasize the importance of employee safety training. This emphasis has been for good reason. Without a doubt, providing routine safety training for workers is one of the most important job duties of the responsible person in charge of safety in any industry. Indeed, most managers know the importance of safety training, but what is not always understood is that specific training requirements are detailed in OSHA regulations. Certain OSHA regulations, for example, state (or in many cases, imply) that the employer is responsible for providing training and knowledge to the worker. Employees must be apprised of all hazards to which they are exposed or have the potential to be exposed, along with relevant symptoms, appropriate emergency treatment, and proper conditions and precautions of safe use or exposure.

In this text we have discussed several OSHA safety and health standards. Employers must comply with these standards and must also require workers to comply. More than one hundred OSHA safety and health standards contain training requirements. Note that although OSHA requires training (OSHA 2006), it does not always specify exactly what is required of the employer or entity that provides the training. Safety professionals must create their own training programs. As it is with other industries, information and instruction on safety and health issues in the food manufacturing workplace are foundational to building a viable organizational safety program. Workers cannot be expected to perform their assigned tasks safely unless they are aware of the hazards or the potential hazards involved with each job assignment.

Hoover et al. (1989) state that in discussions with safety professionals over the years, when they ask, "If you could only keep one part of your activities, which would you keep?" the answer most often given is "Training." Certainly training is important, but even more important is a well-thought-out program, one that is well balanced, aggressive, all-encompassing, that provides continuity, and that gives management the best return for the dollar (is value-based). Safety and other health professionals play a key role in ensuring that all employees at all levels receive the appropriate types and amounts of safety training. Simply, the food manufacturing safety professional must play an active role in preparing, presenting, arranging for the application of, and evaluating safety and health training.

In this section, we cover many of those elements required to conceive and implement an effective organizational safety and health training program. Specifically, we discuss the requirements for a written training program, the need to conduct training, record-keeping requirements, and the need to evaluate the organization training program to ensure that it is both current and effective.

### Written Training Program

In our litigious society it absolutely cannot be overstated how important it is that the safety person ensure that just about anything and everything he or she does or says be in writing. The fact is that OSHA, for one, requires most compliance efforts to be in writing. For example, we point out that the organization's training program must be a "written program" and not just a set of nebulous concepts lodged in the trainer's brain cells.

What should the written safety and health training program consist of? Safety organizations such as the National Safety Council and practicing safety professionals generally cite six basic steps for developing a training program. These are

1. Identifying training needs
2. Formulating training objectives
3. Gathering materials and developing course outlines
4. Selecting training methods and techniques
5. Conducting training sessions
6. Evaluating the training program

#### Identifying Training Needs

As mentioned, more than one hundred OSHA (and other regulatory agencies) standards and regulations have training requirements. Let's take a closer look at what OSHA requires. The OSH Act of 1970 mandates that employers provide health and safety training. It specifically requires

- education and training programs for employees
- establishment and maintenance of proper working conditions and precautions
- provision of information about all hazards employees will be exposed to on the job
- provision of information about the symptoms of exposure to toxic chemicals and other substances that may be present in the workplace
- provision of information about emergency treatment procedures

Simply put, the legal reasons for training are quite clear. Moreover, the safety person should use OSHA's guidance on training, and OSHA's training requirements are a good place to start in formulating the organization's safety program. You should review the OSHA training requirements and make a determination as to which standards or regulations apply to your organization. For example, if the organization's workers are required to wear respirators in the normal performance of their duties, then OSHA's 29 CFR 1910.134 Respiratory Protection Standard certainly comes into play. In reviewing 1910.134, in particular 1910.134(b)(3), it is quite clear that training is required: "The user shall be instructed and trained in the proper use of respirators and their limitations."

Keep in mind that the safety person not only needs to determine what type of training is required for the organization, but also needs to determine what training is not required. For example, under 29 CFR 1910.138 Powered Industrial Trucks Standard, OSHA requires forklift operators to be trained. Specifically, the standard states, "Only trained and authorized operators shall be permitted to operate a powered industrial truck (forklift and other type powered trucks). Methods shall be devised to train operators in the safe operation of powered industrial trucks." Obviously, if the safety person determines that his or her organization has forktrucks, rents forktrucks, borrows forktrucks, and for any reason requires workers to operate such machines, then training is required. However, if the organization does not require its workers to operate forktrucks or other powered trucks, training is not required. Training is expensive, so make sure the training you provide is relevant.

Thus, if a potential hazard is present in the facility for which OSHA specifically requires training, developing and conducting training for the particular hazard should be a top priority. Basically, what it comes down to is that the safety person needs to tailor the organization's safety and health training program to fit the needs of the organization, and to ensure compliance with the regulations.

OSHA's general industry standards (including their training requirements) are numerous and often quite complex. With this in mind, an attempt is made in table 10.1 to simplify the process by identifying typical industry training requirements. Note, however, that table 10.1 is not all-inclusive, and that the frequency of training is not listed.

In addition to the required training listed in table 10.1, the safety person should consider integrating several other topics into the safety and health training program. These include heat and cold stress, environmental rules and regulations, ergonomics, first aid/CPR (29 CFR 1910.151), workplace violence prevention, and other general occupational safety and health topics.

Safety and health training should also be focused on any new equipment or new process introduced into the workplace. Until they have been properly trained, expecting workers to automatically operate new equipment or processes safely is unrealistic. Any time any regulation or company standard operating procedure is modified,

**Table 10.1.    Selected OSHA standards that require training**

| OSHA Standard |
| --- |
| 1.    Hazard Communication Standard (1910.1200) |
| 2.    Employee Emergency Plans and Fire Prevention Plans (1910.38) |
| 3.    Powered Platforms for Building Maintenance (1910.66) |
| 4.    Confined Space Entry (1910.146) |
| 5.    Lockout/Tagout (1910.147) |
| 6.    Respiratory Protection (1910.134) |
| 7.    Hearing Conservation (1910.95) |
| 8.    Flammable and Combustible Liquids (1910.106) |
| 9.    Ventilation (1910.94) |
| 10.   Explosives and Blasting Agents (1910.109) |
| 11.   Ionizing Radiation (1910.96) |
| 12.   Hazardous Waste Operations and Emergency Response—HAZWOPER—(1910.120) |
| 13.   Specifications for Accident Prevention Signs and Tags (1910.145) |
| 14.   Welding, Cutting, and Brazing (1910.252) |
| 15.   Machine Guarding (1910.217–.218) |
| 16.   Powered Industrial Trucks (1910.178) |
| 17.   Fire Brigades (1910.156) |
| 18.   Portable Fire Extinguishers (1910.157) |
| 19.   Blood-borne Pathogens (1910.1030) |
| 20.   Fire Extinguishing Systems (1910.160) |
| 21.   Personal Protective Equipment (1910.132–.138) |
| 22.   Electrical Safety (1910.301–.339) |
| 23.   Excavation Safety (29 CFR 1926.650–.652) |
| 24.   Process Safety Management (29 CFR 1910.119) |
| 25.   Storage and Handling of Liquefied Petroleum Gases (1910.110) |

changed, or revised for any reason, training or refresher training should be conducted. This is particularly the case anytime a new hazardous material is introduced.

The safety person should also look at past performance to determine if a need for additional training exists. For example, if employees have been trained on how to perform a job function in a safe manner, but injury reports and/or equipment failures caused by employee errors increase, then obviously improving employee performance is a definite need. Administering refresher training best effects this.

The safety person should also look at the individual processes and individual job requirements. If a certain process is highly advanced and very difficult to operate correctly (and safely), and if the workers must remember very difficult information, refresher training on an ongoing basis is called for.

*Formulating Training Objectives*

Once training needs have been identified, quantifying them through goals and objectives is necessary. What will the workers gain by completing the training? Effective, clearly stated goals and objectives specifically describe what workers are expected to do, to do better, or to discontinue doing after completion of the training. Goals and objectives are used to determine if workers have attained the desired level of proficiency, and also help the safety professional to determine or measure the cost effectiveness of the program. Goals and objectives, like the training program, should be in writing and presented to each trainee so he or she clearly understands what the training entails and what the trainee is required to learn.

Training works best when

- Workers are given clear learning objectives.
- Facts are presented in a logical sequence.
- Examples come from worker experience.
- The worker can participate and respond.
- Materials are presented in a variety of formats and teaching methods.
- Testing provides direct feedback on progress.
- Built-in repetition provides "imprintation," to delay forgetting.

### Gathering Materials and Developing Course Outlines

Communicating to the trainees the importance of a formal training program is important. This is best accomplished by using a syllabus. Each formal safety and health training program to be presented to workers should have a course syllabus containing the title, course objectives, course outline, evaluation, teaching methodology, and references. The syllabus conveys to the trainee the course expectations, the knowledge to be acquired, the work to be accomplished, and evaluation criteria. Obviously, the syllabus should be developed before the training program begins and handed out before the class begins. (Presently, general safety training practice is to present course descriptions in the introductory section of PowerPoint presentations.)

### Selecting Training Methods and Techniques

Selecting or developing training methods, materials, and techniques is important. During this process, the safety person should keep in mind that the training methods and techniques used should simulate the actual job function as closely as possible—training custodians on rocket science makes little practical sense. The actual training methods employed depend on the material, the size of the class, and the instructor's style.

### Conducting Training Sessions

In determining what is to be covered in each training session, instructors should also decide what teaching method to employ, approximately how much time it will take, what visual aids or demonstrations to use, and what materials must be ready to hand out to workers. In our experience, when training sessions include the use of demonstrations and other visual methods, the training is more acceptable to the

**Course Evaluation Form**

Program Evaluation

Quality Improvement

Program Title:

Program Date:

Instructor:
Thank you for participating in the training program. We hope that the presentation will enhance your job skills and further your professional development. We strive to achieve a higher level of quality in training to meet the needs of operations and to achieve increased safety awareness for all employees. To meet our needs now and in the future, we would appreciate your input and answers to a few questions about the subject material and instructional quality of this training session.

Please check the box under the number that best reflects your evaluation for the following topics (1 = poor, 5 = excellent, n/a = not applicable).

| TOPICS | 1 | 2 | 3 | 4 | 5 | n/a |
|---|---|---|---|---|---|---|
| Correlation between the program and the objective | | | | | | |
| Safety Management | | | | | | |
| Operations Information | | | | | | |
| Motor Vehicle Operations | | | | | | |
| Construction Safety | | | | | | |
| Hazard Communication | | | | | | |
| Air Monitoring | | | | | | |
| Lockout/Tagout | | | | | | |
| Personal Protection | | | | | | |
| Site Safety Plans | | | | | | |

Please answer the following questions as completely as you can. Use the back of the form for additional space if needed.

1) What was the best feature of the training session?

2) Do you have any suggestions for new sessions?

3) Do you have any additional comments on how we can improve the program?

Adapted from Roughton and Whiting (2000).

FIGURE 10.1
Course evaluation form.

workers—they buy in on the presentation, which means they are likely to learn much more than they otherwise would.

### Evaluating the Training Program

The primary way the effectiveness of the training program is determined is by evaluating it. In fact, the most critical part of the training process is that of evaluation. Training programs are typically evaluated based on results. Asking questions, both in formal testing and in surveying the trainees, can give the presenter valuable information on how to make training most effective. Did the training correct the deficiency? Did the training supply the information the workers need to perform their assigned duties in a safe, productive fashion? Evaluation is all about feedback. Feedback is obtained by asking the questions above, and by asking whether the training was helpful. Did the workers get what they wanted? Was there anything that they would have liked done differently? It should be apparent that the answers to these questions can best be obtained by asking workers who have been trained. Using a formal preprinted course evaluation form (see figure 10.1) is a practical, proven technique. Those parts of the program receiving poor evaluations should be revised promptly.

### Record Keeping

We have stated previously that no matter how much training an organization provides, no matter how detailed it is, no matter its quality and content, and no matter how effective it is, if the training is not properly documented, the regulators and/or courts of law will consider that the training *has not* been done.

Beyond the regulatory and legal requirements of documenting training, documentation is also required to provide a record for the manager to review to determine who has and who has not been trained. Obviously, this is important. Additionally, many training sessions must be routinely repeated (refresher training) on a periodic basis. Keeping up-to-date and accurate records helps in this process.

## REFERENCES AND RECOMMENDED READING

Hoover, R. L., R. L. Hancock, K. L. Hylton, O. B. Dickerson, and G. E. Harris. 1989. *Health, Safety, and Environmental Control.* New York: Van Nostrand Reinhold.

LaBar, G. 1991. Worker Training: An Investment in Safety. *Occupational Hazards*, August, 25.

Laing, P. M., ed. 1991. *Supervisor's Training Manual.* 7th ed. Chicago: 35.

Occupational Safety and Health Administration (OSHA). 2005. *Small Business Handbook.* www.osha.gov/Publications/smallbusiness/small-business.html.

———. 2006. All about OSHA. www.osha.gov/Publications/all_about_OSHA.pdf.

———. 2008. elaws—OSHA Hazard Awareness Advisor. www.dol.gov/elaws/osha/hazard/ keywords.asp.

Parachin, V. 1990. 10 Tips for Powerful Presentations. *Training*, July/August, 71–83.

Roughton, J., and N. E. Whiting. 2000. *Safety Training Basics: A Handbook for Safety Training Program Development*. Rockville, MD: Government Institutes.

Spellman, F. R. 1996. *Safe Work Practices for Wastewater Treatment Plants*. Lancaster, PA: Technomic.

# Appendix A

## Programas de OSHA para Ayudar al Trabajador Hispano

La misión de la Administración de Seguridad y Salud Ocupacional (OSHA, por sus siglas en inglés) es la de velar por la salud y seguridad de los trabajadores, a través de la implantación y vigilancia del cumplimiento de estándares y la provisión de entrenamiento, educación e información; estableciendo alianzas y alentando las mejoras permanentes a las condiciones de seguridad industrial y salubridad del lugar de trabajo.

En el año 2001, OSHA conformó la "*Hispanic Task Force*," una fuerza especial que continua reuniéndose de forma regular para identificar las formas de reducir las averías, enfermedades y accidentes mortales entre los trabajadores y empleadores de habla hispana.

### COMUNICACIÓN

Durante el 2002, OSHA ha creado una pagina en español en su sitio de información en Internet, también ha agregado una opción en español en su línea gratuita de ayuda en el 1-800-321-OSHA. Por otro lado, OSHA revisó y encontró entre su propio personal, 180 personas del nivel federal y estatal que hablan español. OSHA ha identificado también a coordinadores regionales hispanos para supervisar las acciones de llegada al trabajador hispano. Los esfuerzos de reclutamiento de trabajadores hispanos continuan en todas las oficinas.

Para mejorar la comunicación con los hispanos, la "*Hispanic Task Force*" recientemente ha creado un glosario de términos con mas de 200 palabras relacionadas a la seguridad y salubridad en el trabajo; este glosario se lo encuentra en el sitio de Internet de OSHA, www.osha.gov. Más de 15 millones de oyentes han escuchado los anuncios públicos de OSHA en 650 radioemisoras en español, haciendo énfasis en la seguridad y la salud en el lugar de trabajo.

## TRADUCCIONES

OSHA cuenta con una docena de publicaciones y diez hojas de datos en español en su sitio de Internet. La idea es continuar agregando material de forma permanente, a medida en que se van traduciendo. También están disponibles en el sitio OSHA en el Internet, dos programas de computadora interactivos, denominados "*eTools*," el usuario puede acceder y bajarlos de la red a su computador de forma gratuita. Uno de los programas trata sobre la costura comercial y el otro sobre la construcción.

## ENTRENAMIENTO Y CAPACITACIÓN

Los 70 especialistas en asistencia de cumplimiento de normas ofrecen talleres y seminarios de capacitación a la medida del trabajador hispano, desde los jardineros y paisajistas de Georgia hasta los trabajadores en madera y carpinteros en Connecticut, incluyendo a los trabajadores temporales en Texas e Illinois. Adicionalmente, OSHA cuenta con 20 centros educativos en 35 ubicaciones distintas, que pueden ofrecer algún tipo de entrenamiento en salud y seguridad en idioma español. Por otra parte, 50 organizaciones sin fines de lucro están recibiendo fondos del gobierno para desarrollar material de entrenamiento para acceder a los trabajadores de habla hispana, por ejemplo:

- La *Georgia Tech Research Corporation* está desarrollando materiales de entrenamiento para enseñar sobre los peligros que existen para los conductores de camiones-mezcladores de concreto.
- La *International Society of Arboriculture* está creando un CD-ROM interactivo acerca de las practicas de seguridad que deben utilizarse en el trabajo de atención a los árboles.
- La *Pennsylvania Foundry Association* está preparando material para enseñar sobre la prevención de la silicosis en la industria de la fundición.
- La *Texas Engineering Extension Services* está desarrollando materiales de entrenamiento con fotografías, sobre los peligros en los campos de operaciones de petróleo y gas.
- La Universidad de Massachusetts, a través de la Fundación de Investigación *Lowell* hará posible, una versión de 10 horas, en español de los cursos del programa en la construcción.

## ESFUERZOS CON LOS ALIADOS

Además de nuestras alianzas con México y los consulados mexicanos, OSHA ha desarrollado 10 alianzas nacionales y 31 alianzas regionales que se esmeran en llegar a los trabajadores hispanos en temas como ergonometría, seguridad en manejo de vehículos motorizados, amputaciones, caídas y seguridad en la zona de trabajo. Como parte de una de las alianzas, OSHA y la asociación Nacional de Constructores de Casas (*National Association of Home Builders*) están desarrollando una hoja de datos en el tema de protección de caídas, en idioma español.

Uno de los participantes de nuestros programas de protección voluntaria, Wenner Bread en Bayport, Nueva York, tienen una de las tasas más bajas de averías y enfermedades en la industria, cerca a la mitad de las demás. La empresa atribuye su éxito al hecho de que realiza todos sus entrenamientos y materiales educativos también en idioma español como parte de un efectivo programa. La empresa realiza reuniones cortas diariamente y reuniones semanales de seguridad en ambos idiomas. La empresa utiliza a los propios empleados bilingües para que interpreten y traduzcan las presentaciones en el lugar de trabajo. Los esfuerzos de la empresa Wenner han conducido también a una mejor relación laboral, mayor productividad y mejor calidad en el producto final.

Los días de entrenamiento en seguridad industrial donde participa la familia, atraen a los trabajadores para su capacitación, mientras, al mismo tiempo proveen diversión y educación para los otros miembros de la familia. En Dallas, al principio de este año, OSHA hizo una alianza con los Contratistas Hispanos y el consulado mexicano para organizar un evento familiar. Alguno empleadores pagaron la asistencia de sus trabajadores si participaban en por lo menos 10 de las 12 sesiones.

El 27 de marzo del 2004, se llevó a cabo la Feria Familiar de Seguridad y Salud, en Hialeah, Florida. La feria fue auspiciada por OSHA, la División de Salario y Horas del Departamento de Trabajo de los Estados Unidos y otros cuatro grupos. La feria presentaba ocho clases sobre el tema de seguridad, todas en español.

## CUMPLIMIENTO DE NORMAS

OSHA está investigando si existe alguna conexión entre las barreras culturales y de idioma y los accidentes fatales de trabajadores hispanos. Los hallazgos preliminares indican que cerca a un 25% de los accidentes fatales están relacionados de alguna manera a barreras culturales o de idioma. La agencia continuará recolectando y analizando datos para determinar formas para eliminar estas barreras y reducir así, los accidentes mortales en el trabajo.

Source: www.dol.gov/opa/media/press/opa/OPA20041371-osha-s.htm

# Index

# About the Authors

**Frank R. Spellman** is assistant professor of environmental health at Old Dominion University, Norfolk, Virginia, and author of 55 books.

Spellman's book titles range from *Concentrated Animal Feeding Operations* (CAFOs) to several topics in all areas of environmental science and occupational health. Many of Spellman's texts are listed on Amazon.com and Barnes & Noble. Several of his texts have been adopted for classroom use at major universities throughout the United States, Canada, Europe, and Russia; two are currently being translated into Spanish for South American markets.

Spellman has been cited in more than 400 publications; serves as a professional expert witness for three law groups and an accident investigator for a northern Virginia law firm; and consults on homeland security vulnerability assessments (VAs) for critical infrastructure, including water/wastewater facilities nationwide.

Spellman receives numerous requests to coauthor with well-recognized experts in several scientific fields. For example, he is a contributing author for the second edition of the prestigious text *The Engineering Handbook.*

Spellman lectures on homeland security and health and safety topics throughout the country and teaches water/wastewater operator short courses at Virginia Tech.

Spellman holds a BA in public administration, a BS in business management, MBA, an MS in environmental engineering, and a PhD in environmental engineering.

**Revonna M. Bieber** holds an AAS in radiography and a BS in biology. She is currently a graduate student at old Dominion University, where she is obtaining an MS and MPH in environmental health and safety. She has expertise in environmental health hazards as well as radiologic and health care safety.